The Planetary Garden AND OTHER WRITINGS

The Planetary Garden AND OTHER WRITINGS

Gilles Clément

Translated by Sandra Morris

Foreword by Gilles A. Tiberghien

PENN

University of Pennsylvania Press

Philadelphia

Penn Studies in Landscape Architecture

John Dixon Hunt, Series Editor

This series is dedicated to the study and promotion of a wide variety of approaches to landscape architecture, with special emphasis on connections between theory and practice. It includes monographs on key topics in history and theory, descriptions of projects by both established and rising designers, translations of major foreign-language texts, anthologies of theoretical and historical writings on classic issues, and critical writing by members of the profession of landscape architecture.

The series was the recipient of the Award of Honor in Communications from the American Society of Landscape Architects, 2006.

"The Planetary Garden: Reconciling Man and Nature" originally appeared as *Le Jardin planétaire: Réconcilier l'homme et la nature* © Éditions Albin Michel, 1999.

"Life, Constantly Inventive: Reflections of a Humanist Ecologist" originally appeared as *Toujours la vie invente: Réflexions d'un écologiste humaniste* © Éditions de l'Aube, 2008.

"The Wisdom of the Gardener" originally appeared as *La Sagesse du jardinier* © Éditions Jean-Claude Béhar, 2006. Published by arrangement with Éditions Jean-Claude Béhar, Paris, France.

Published by
University of Pennsylvania Press
Philadelphia, Pennsylvania 19104-4112
www.upenn.edu/pennpress

Printed in the United States of America on acid-free paper
10 9 8 7 6 5 4 3 2 1

Cataloging-in-Publication Data is available from the Library of Congress
ISBN 978-0-8122-4712-1

Frontispiece: Gilles Clément

CONTENTS

DELVING INTO THE CONCEPT

Gilles Clément, horticultural engineer, entomologist, landscape architect, and writer, occupies a special place in French professional circles that have long been dominated by an urbanistic vision of the role of the landscape architect. By championing in the 1970s a particular vision of the garden and nature, at a time when landscape was more readily used in the service of grand social utopias, he was somewhat marginalized. His completed projects, such as the Parc André-Citroën in Paris, in collaboration with Alain Provost and Patrick Berger from 1985–1992, the Parc Henri Matisse, opened in 2003 at Euralille with Eric Berlin and Sylvain Flippo, and the same year the garden of the Musée du Quai Branly in Paris with Jean Nouvel, introduced him to a wider public and established him among his peers as one of the most important contemporary landscape architects. As a writer and talented lecturer—in 2011–2012 he held the Chaire Annuelle de Création Artistique at the Collège de France[1] and was in demand everywhere abroad—he has popularized the concepts that he has invented, and that have helped him to design his gardens in the context of an ecological view of man's place in our world.

Alain Roger was right when he wrote, in an essay that he devoted to Gilles Clément, "Du jardin en mouvement au jardin planétaire," that "if it is true that philosophy, as Gilles Deleuze once stressed, is characterized by the creation of concepts, then Gilles Clément certainly deserves the title of philosopher."[2] Beginning with experiments in his garden laboratory, La Vallée, located in the Creuse, one of France's least populated departments, Clément, with the idea

---

[1] The inaugural lecture was published under the title *Jardins, paysage et génie naturel* (Paris: Éditions Fayard, 2012).

[2] Extracts from this text are published as an appendix to the second edition of Clément, *Le Jardin en mouvement* (Paris: Sens & Tonka, 2001).

of the "garden in movement" as his starting point, has developed a series of concepts central to his way of thinking, the most important of which can be considered here.

The Garden in Movement

Clément set out this idea for the first time in an article published in 1984 under the title "la friche apprivoisée,"[3] then in 1991 as "Jardin en mouvement," the title of his book. For the gardener, this means a certain way of promoting biodiversity through an intelligent management of natural spaces by relinquishing the control of plants, to which gardening is too often reduced.

The garden in movement, Clément writes on his web site, "takes its name from the physical movement of plant species across the plot, which the gardener may interpret as he sees fit. Flowers germinating as part of a transitional process present the gardener with a choice of deciding whether he wants to preserve the transition or preserve the flowers. The garden in movement advocates preserving the species having taken a decision on their chosen position." This principle and the practice that it implies were developed in his book *Eloge des vagabondes* (2002).

The Planetary Garden

The idea of a planetary garden, *le jardin planétaire*, conceived more than fifteen years ago and already present in *Thomas et le voyageur* (1998), is succinctly expressed in *Toujours la Vie invente* (see "Life, Constantly Inventive" in this volume). A "garden," Clément reminds us, "signifies enclosure." Therefore, to speak of a garden on the scale of the planet seems paradoxical, because in this case the enclosure has disappeared. But not at all, it has simply become considerably

[3] Clément uses the word *friche* to describe land that has been left uncultivated for a number of years, and is systematically invaded by a series of vegetation. In this case, the natural progression is "tamed" (*apprivoisée*) by human intervention.

larger: "Instead of being limited to a small space that we control, from now on the garden is placed within the limits of the biosphere. This is the new enclosure." And this garden is entrusted to our care. Hence, a feeling of collective responsibility: what we do here will inevitably have repercussions over there, on the other side of the planet, to the extent that each one of us, in our daily activities, in our way of understanding the world and transforming it, however little that may be, can in our own way be a "planetary gardener."

The Third Landscape

The third important concept is that of the "third landscape," *le tiers paysage,* which Clément articulated shortly after the exhibition *Le Jardin planétaire* in the former market of La Villette, Paris, in 1999–2000. The idea was prompted by an artistic commission offered to him by Guy Tortosa, then director of the Center of Art and Landscape at Vassivière in 2002. The result was an essay, published first as *Manifeste du tiers paysage* (2004), then taken up again and developed under the aegis of the Collectif du Chomet, a group of landscape professionals (2012). This notion of a third landscape, which borrows its name from the "third estate," a term coined by the Abbé Sieyès during the French Revolution to define those who were neither the nobility nor the clergy, identifies in this case "the totality of all those places abandoned by man." In this essay, Clément sets out from the precept that between the management of the forests and that of the fields there exists an edge condition outside the jurisdiction of forestry or agricultural engineers, an intermediate space that supports a biodiversity not found elsewhere. Subsequently, he extends this observation to other places, such as road shoulders, riverbanks, moors, industrial wastelands, and so on. It is to these ecological reserves that he draws the attention of ordinary citizens, but also that of the decision makers on matters of urbanism, in order to ensure the preservation of these areas and to make them a mainspring for action and thought for future societies. "The third landscape" is "a special place of biological intelligence," he declares, at the same time characterizing it as having an "aptitude for constant reinvention."[4]

A Humanist Ecology

In France, Gilles Clément is seen as part of a family of contemporary thinkers concerned with the natural world who, like Francis Hallé,[5] Pierre Rhabi,[6] and Pierre Lieutaghi,[7] have faith in the resources of the planet and in the capacity of human reason to call a halt to its own folly. Hence their perseverance, their militant activities as lecturers and writers, in the service of this conviction, however different they may be. Ecology in its broadest sense as the balance between man and nature is their credo. It is what Gilles Clément, together with Louisa Jones, in a book that came out in 2005, called a *humanist ecology*. This term, already introduced in the exhibition *Le Jardin planétaire* at La Villette in 1999–2000, opposes that of a "radical ecology, according to which life on earth can exist without man." Now even if this point is indisputable, Clément says, we have nevertheless the right to ask ourselves, "Who would there be to appreciate radical ecology if man actually disappeared from the planet?"

Gilles Clément's house at La Vallée is really a hut.[8] He built it without any preliminary plan using materials found on site. It discovered

---

[4] Gilles Clément, *Manifeste du tiers paysage*, an enhanced version of "Evolution et mise en pratique du concept de tiers paysage" (Entremont: Tiers Paysagiste Press, Le Collectif du Chomet, 2012), 93.

[5] Botanist and biologist specializing in the ecology of humid tropical forests. From 1996–2023 he directed the scientific missions of the treetop raft in the canopy of tropical forests. See the letter addressed to him by Gilles Clément in *Où en est l'herbe? Reflexions sur le jardin planétaire*, texts presented by Louisa Jones (Paris: Actes Sud, 2006), 106.

[6] Agriculturist and biologist, essayist and novelist, author of among other works *Manifeste pour la terre et l'humanisme: Pour une insurrection des consciences* (Arles: Actes Sud, 2008).

[7] Ethnobotanist, author of numerous works on the relationships between plants and society.

[8] These notes are a continuation of an initial series of five notes written on different occasions and published in *Notes sur la nature, la cabane et quelques autres choses* (Paris: Éditions du Félin, 2008). Although each essay is independent, they also complete and respond to one another.

its form through the stones and beams that he was able to lay his hands on to put it together. Its overhanging location lends it the air of a lookout, but it is almost as if it had been placed there to allow a better view of the passing birds and the animals making their way down to the lake. In the morning, the eastern side of the building is bathed in sunlight from nine o'clock. It is a low-angled golden light, made even softer by the washed-out gray wooden shingles of the roof that slopes down almost as far as the terrace, only coming to a halt some fifty centimeters from the flagstones.

From the window where I am writing, facing south, I have just seen a deer passing the edge of the wood defining the upper boundary of La Vallée. It seems to me to be on the outposts of the "civilized" world, on the frontier between my world and that of nature, both equally illusory if considered separately. The animal did not come back in my direction; it simply slipped off to the right and then disappeared. I caught sight of another deer two days ago, while walking on the other side of the wood. But it passed so quickly that it took me a while to realize what it was.

Gilles Clément had said to me, "I live with animals," and I realize that that is absolutely true. He wrote incidentally in *Le Salon des berces:* "At La Vallée the animals inhabit the garden. Hence, the house. In their way, they define the rules of a territory that we, humans, merely pass through." The same idea can be found in Rousseau, in *La Nouvelle Héloïse,* where Julie says, when she is showing her garden to Saint Preux, amazed to observe that the birds in the aviary are treated like guests and not like prisoners: "What do you mean, guests? . . . It is we who are the guests; they are the masters here, and we pay them tribute for occasionally being tolerated by them."

The lizards, the frog, the one that I see every day when I go down the thirteen steps of the terrace, and that jumps into the water just as I pass near its pond, the rabbits that scatter when I make my way toward the bottom of the garden, the dormice that come out at night around eleven o'clock, and go back into the house in the morning around five, making a hell of a racket, these are the masters of the place.

Today the water was still a little cold when I swam in the lake. On the opposite bank, there were a few people enjoying themselves and two motor boats passed in front of me. All in all, heavy traffic for this corner of the world. But it is Sunday. Generally, if you see people, it's only one or two couples on the little beach on the other side. Apart from this kind of event, which demonstrates the presence of humans on this earth, you see no one, and the bed of the Creuse lies hidden from sight between its steep banks.

At the end of *Salon des berces,* Clément describes the party that marked the completion of the construction, celebrating the community, thanking the participants, and opening up the future of a place destined for the meeting of friends and endless entertainments. They were to perform *Les Mamelles de Tirésias;* we are going to "feel free to take every liberty with it, roll Apollinaire in the Creuse fondue, he'll never recover," Clément wrote. The performance began on the terrace and continued through the different garden "rooms." "That evening the intention was to celebrate the ordinary. The everyday extravagance of nature, the everyday imaginative power of our spirits continually struggling against the rules of conformity. The ordinariness of human beings driven by their desires and not by ulterior motives." A century and a half later, on a totally different continent, and without necessarily being aware of it, Clément was reinvoking the ideas of Thoreau and Emerson, together with the collective and transcendentalist utopias of Fruitlands and Brook Farm, but without following any religious principle. Another such idea is planetary intermingling, a phenomenon that is as much physical as mental. The *pollen* of thought, so dear to Novalis (whose pseudonym means "a clearer of land," hence *friche*) is dispersed throughout the world, and here and there ideas reemerge spontaneously, taking on new colors, ignoring their relationship with earlier ideas that were formulated in other times and places, which resemble them.

Every evening, around half past six, I water the vegetable garden, its six square beds where zucchini, tomatoes, potatoes, beans, and carrots grow, but also leeks, salads of different kinds and all sorts of herbs for the kitchen, of which chives are my favorite. In one of the beds, orange marigolds and violet-blue borage form a marvelous marriage of colors, hiding the lettuces and beetroots that grow beneath them. Growing in the corner are several varieties of verbascum, their yellow flowers reaching more than two meters in height. The other evening I gathered a dozen potatoes. "Gather" is not perhaps the right word. Rather, I burrowed in the soil to feel the tubers, hard and round under my fingers, unearthing them like turtle eggs buried in the sand. I had noticed one with its pinkish top sticking out, but I had to search deeper for the others and I know that there are still many more nearby. It's a strange pleasure to find them like this, fully formed, nourished by the earth and still in their element. Eating them afterward adds an entirely different imaginary and sensual dimension to the act. I discover that this fine surface which has the appearance of a carpet of vegetation is actually a covering of shredded grasses—mulch—destined to regenerate and protect the soil.

When you go to swim, you follow the path that leads to the little creek below La Vallée. Having reached the bank, you can go down into the water by a natural flight of steps on which you leave your things before following this staircase that disappears beneath the dark surface of the lake. Then you plunge in and, struck by the cold, move about a bit until your body is sufficiently warmed up for you to feel like shouting "It's great!" But I'm alone and have nobody with whom to share this feeling. Not far from this little natural quay, some students, visiting a few months ago, helped Clément to open up the view of the Creuse from the top. They had built two small stone cairns that immediately caught my attention. I decided to add another, as a tribute to my host, and by way of a friendly gesture to artists like Hamish Fulton, Richard Long, and Chris Drury, who have left so much in the landscape.

Clément explained to me which road to take to the village when he came to pick me up. I have the use of his Peugeot 104, which looks ten years older than its thirty-five years, the gear lever reduced to an iron rod that one is afraid of embedding in the palm of one's hand when using it, the left-hand windshield boarded up, the front passenger seat exposing the brownish foam of its stuffing, the car generally stout-hearted and willing, capable of facing hills of forty-five-degree incline—like the one you have to take when you emerge from the woods to join the road that passes the house of the neighbor, Nadine. On the way back, you leave the national highway at the signpost indicating the boundary between Indre and Creuse and enter into a maze of little roads that I am gradually getting to know. Very soon after this crossing, you pass a modern house surrounded by land enclosed with a wire fence. It's the house of the "terrorist," as Clément calls him. Its owner, jealous of his plot of land and no doubt fearing that someone might dare to place an illegal foot on it, has diligently proscribed a space without actually hiding it from the eye, ostensibly revealing his orchard with its military alignment to everyone's envy and satisfaction. "I have good tobacco in my pouch, I have good tobacco but it is not for you," he seems to be telling us, humming that ancient air.

What does this man look like, whom I have never seen? I imagine him with heavy eyebrows and a thick moustache, short-legged, his head sunk deep into his shoulders, shooting a look of distrust at the person speaking to him. But perhaps he is a totally different kind of man, the managerial type, neat and freshly shaven, with the fascist and arrogant tendencies of the self-satisfied pseudo-countryman.

The hut, as I have said in another note, has neither interior nor exterior, it is completely open to nature by which it is somehow permeated. On coming across the following passage in *Salon des berces*, I found the same idea and understood better why Clément, having read my piece, told me that he too had built a hut.

> Essentially, one day an architecture discovers its limitations. At La Vallée, the inhabited space comprises the parcel of land in its entirety and not the restricted surface of the building. It is my hope that beyond the built, and at whatever moment that may have reached its limits, the project, instead of coming to a conclusion, will continue to expand, and the garden will lead the eye forward indefinitely. . . . Through the twenty-one openings in the house, though small, the garden springs up on every side. The interior is only a space through which the eye travels. The view stops at the oaks, the hornbeams, and all those layered trees: the chestnuts in the distance. Without having noticed, I find myself head of a house of five hectares, of which only a hundred square meters are covered by a roof.

This morning, I spent some time looking closely at the frog, Gwendoline, approaching her quietly so as not to frighten her. I had already surprised her once on her wooden plank, where she is always on the lookout ready to jump on the first insect to appear, or to leap into the water as soon as her environment is disturbed by my presence. She is a very fine specimen, sporting a line of deep green down the middle of her back, making her look like some sort of paperweight for a rich connoisseur of frogs and toads. You might almost think she was carved from a dark glassy stone, with a central vein of sparkling chlorophyll. Not far away on the terrace, as soon as the sun appears, I notice the lizard with the cut-off tail that I have named Boniface. He comes and goes, with his apparently unsteady gait, an offbeat rhythm to his step, his head held in an ultrasensitive position like radar for detecting gnats.

I observe my companions with an attention that I only rarely give to their species, and each day that goes by increases my curiosity about them. I am immersed in nature, in a garden created by a man whose dream I share through a secret door by which he has allowed me to enter to pursue my own dream. A garden, Clément says in "Life, Constantly Inventive," is a "territory where everything is intermingled: flowers, fruit, vegetables," adding, "I define the garden as the only territory where man and nature meet, in which dreaming is allowed. It is in this space that man can be in a utopia that is the happiness of his dreams."

In the library here, I find lots of books on mushrooms, flowers, insects, and birds. The large volumes of Adalbert Seitz on *Les Macrolépidoptères du globe* are particularly fascinating, printed in French in Stuttgart exactly a century ago—the German original dates from 1909—in full color with detailed and very elegant reproductions. At the time, Seitz was director of the Zoological Gardens in Frankfurt. The book begins: "Thanks to their brilliant and pleasing appearance, *lepidoptera,* and to a lesser degree other insects, form such an essential part of the character of a summer landscape that one can scarcely imagine any country blessed with a rich vegetation deprived of this element of life." A fine lead-in by a scientist who goes on to tell us that butterflies have not always existed, and that in all likelihood they appeared on our planet at more or less the same time as humans. The illustrations are splendid, and the butterflies are arranged in rows, sometimes with only half the body represented and one open wing. La Vallée is full of butterflies, and when Clément came here as a child, it was precisely to look for them. Even then it was called "the valley of butterflies."

Childhood offers us our first profound experiences of space and sometimes of nature. What happens then remains with us all our lives. Some people will remember a particular smell, others sounds, or a certain quality of light. Whatever they may be, these sensations will always be attached to specific places, to specific spaces where in some way we have deposited them and which serve to preserve them, even if no one but we ourselves may be aware of it. Accordingly, beneath the real landscapes, these layers of sensory recollections define other landscapes, those of our memory whose forms we perceive circulating slowly like those vast flattened capsules of air that are displaced in winter beneath the ice of frozen ponds. Childhood is our prime field of experience, and in Clément's case, it was the garden that generated the hut and not vice versa, it is the garden that somehow gives it its meaning. "When I bought this place to live here," he wrote in a book of essays, *Nous réconcilier avec la terre*, "I approached it with the eyes of a child. I became a small child again. My project was not to build a house there with a garden around it. It was in fact the opposite: I wanted to live in a garden."

On the beam that supports the top of the window of Clément's studio, on the east side of the house, someone has engraved in the wood a circle with an arrow beside it, and further away a bird with its wings outspread. The meaning of this design escapes me. Clément told me that it was very difficult to understand. Something like the circle of time enabling us to approach our fundamental aspirations? No, that's not right. Something to do with the eternal cycle? I doubt it. In fact, when he came back, Clément explained it to me: this circle, which is not closed, represents creation evolving, which leads to the peace symbolized by the dove. It can be summed up as "creating for peace."

A torrential rain is beating down on La Vallée. The water streams over the shingles and flows through the stone channels away from the terrace and into the little basin that lies under the slope of the roof, just in front of the desk in the studio, which is separated from the storm by only a broad pane of glass. It gives me the feeling of being outside, my eyes saturated with the light filtered by the grayness of the sky that extends above me, dull and joyless. Nothing melancholy about it, though. It gives essentially the impression of an overspill pouring out over the world like a wineskin that is overflowing. It is the countdown of time that is beginning, the rhythm produced by the impact of the raindrops on the soil, the vertical grid that they form, practically imperceptible, making it almost tangible before overwhelming me, delivering me up to the fluidity of the seconds, minutes, and hours of the day. It is the "time of the garden," for me a fresh experience here each day, because as Clément says in *Une Brève Histoire du jardin*, it is a time that "is constructed from day to day." And then, precisely, to wonder: "Who today has the time to take that time? Who allows himself to forget the passing of time to appreciate time as it occurs? Who is sufficiently concerned with the living material of a growing garden to be satisfied with it the way it is and follow it in its evolution?"

Today, I happen to open Thoreau's journal at 15 March 1842 and come across this: "Nature is constantly original, and endlessly invents numerous models like a craftsman in his studio." Corresponding perfectly with the title of Clément's book *Toujours la Vie invente.* At that moment I am struck by a certain number of comparisons that it is possible to make between Clément and Thoreau. Among them is this reciprocal desire to let go, to give nature a free rein by abiding by the principle that Clément defines as "working with nature, never against her." This is without counting what they share, a genuine political engagement, in its most fundamental sense.

In the hollow of the valley, not far from the house, the katsura, which is called the "caramel tree" because it smells of caramel in the spring, was planted by Clément when he was building the house. Since then it has reached some twenty meters. Just beside it, near the stream of La Berthonnière, grow the magnificent and huge Chilean gunneras, sometimes called giant rhubarb, which enjoy humidity. Further up the slope, you find the evening primroses again, those extraordinary brilliant yellow flowers that open at night and fade in the morning when the sun warms up. Every day, the flowers that opened the previous evening die, and every evening others are born and open before the eyes of anyone who takes the time to observe them, just as in those scientific films that accelerate the growth of plants. In his *Eloge des vagabondes,* Clément writes that there are two species that travel through his garden, *Oenothera biennis* and *Oenothera erythrosepala*, known as Lamarck's evening primrose (or redsepal evening primrose), which gives him an opportunity to return to the topic of transformism, of which he is particularly fond. His book *Nuages* (2004) is profoundly Lamarckian, and at the abbey of Valloires in the Somme he has created a garden in homage to Lamarck. Clément never misses an opportunity to oppose this theory to that of Darwin's natural selection, which appeared fifty years later. Darwin, Clément writes, proposed "a system of evolution through selective pressure: the strongest wins. The strongest meaning the best adapted to a specific environment, with this worrying corollary; any form adapted to a specific environment is destined to disappear as soon as the living conditions of that environment change. Unless it transforms itself . . . which brings us back to the Lamarckian viewpoint."

In the absence of its owner, I lived in this place as if it were mine. A feeling that no doubt corresponds with the way in which the place was originally conceived. All Clément's concepts speak about nature as well as about humanity; they evoke a possible community of humans and nonhumans, a way of constantly inventing new forms for living better together. The term "humanist ecology" says the same thing. In that sense, La Vallée is not only a small laboratory for observing the spontaneous movement of plants and their planetary intermingling, but also a place for friendship and sharing. A base behind the lines from which to think differently about the world and to act accordingly.

The Planetary Garden

RECONCILING MAN AND NATURE

Acacia fleur
la Villette Mai 09
Gilles Clément

Le Jardin planétaire *(The Planetary Garden)*
exhibition, Parc de La Villette, 2000.

The guide we had chosen was surly, with a slight stoop. He looked like a bear in costume, his white shirt fastened at the neck without a tie, one corner of the collar somehow raised toward the sky, the other hidden under a black jacket that seemed to be asymmetric. He was nodding gently and smoking, belching huge oblique puffs of smoke. Meanwhile, people were moving toward him. We waited for the clouds and persistent smell of tobacco—officially prohibited here—to disperse before approaching him to listen.

We had come a long way. Having set out the previous day from an ordinary garden somewhere in the country, our journey had ended in town at the doors of the "planetary garden." Just a brief pause. From now on we would rely on words to transport us; approaching the world from a new angle, the TGV, sharpening our wits, meant leaving behind us between stations fragments of history passed through all too quickly. Seemingly, we were going to speed up the pace between past and future within a few steps. We were ready. Inside our bags, some ham and bread, and for Lucien, we knew, a flask of wine.

During the previous week there had been growing excitement in the village. Another Parisian folly, a treat for the rich. The weather forecast made the decision for us, it was cold, the flooded marsh-lands were beginning to freeze, no possibility of working in the vegetable garden any longer. A small group was organized to get refundable tickets. We would also have to take the métro.

We talked about it that evening. Lucien the sailor knew the Mediterranean, his own personal sea. He loved that image, the waves, the horizon. It's there, do you see, the tiny Mediterranean in the blackness of space: he was showing us the planet. The others told him, you're dreaming, and he replied "yes." We tried being serious, laughing; but it was he who led us on. You have to know how to respond to children—Lucien was beyond that age, but not to us—when they try one last time to fend off the night, begging us: please, bear, tell me a story.

ENDEMISM

"Endemism" is diversity through isolation, a diversity of creatures and of ideas. Geographic isolation and climatic barriers create as many environments where species appear.

The more habitats (biotopes) there are, the more species there will be capable of living there and societies capable of developing.

The longer the habitats remain isolated from one another, the longer diversity remains. It is expressed through the variety of individuals, behavior, and beliefs.

The Carousel of the Continents

No doubt that was why, from the very beginning and without any discussion, we had chosen this man, standing on his hind legs, stuffed into a black suit shiny with wear like the fur of old grizzly bears, who seemed to be expecting us.

You're just in time, I was about to begin. Whereupon, he tied up his laces, grunting. We had made the right choice.

Our journey on the merry-go-round began with a delicate dance set against the oily background of the magma. The continents were moving. A slow evolution, reversals, couplings, retreats. The elasticity of the skin, the folds and mountains, abysses, plains and smooth surfaces, oceans; nothing remains in place. We have now reached the third Pangaea. At least that is the thesis proposed by contemporary scholars.

Carried along with the flow, we watched the past deconstructed, the present composed of fragments, and already the outline of the future. We used to think of the Earth as stable, firm and compact, but there it was, unbridled, flexible, inventive, activated by the energy of a formidable cosmic machine.

You'll see, said our guide, it's a game, life has its rules, we enjoy it, particularly (and here he suddenly had a hungry look) when the rules change.

Change, a new orientation, from the Cambrian to the present day—a minute fragment of earth's history—the only thing to do, to try it, to erase it, to begin again. That applies not only to the shape of the planet but also to the creatures that inhabit it. The form of the continents as they were several million years ago has nothing to do with that of today, and practically no species survives from those past eras. Everything we know is new, quaternary and perishable. When we reach the quinternary, the oceans will have different contours, as will the continents, living creatures will have different forms, different names. For sure, there will always be simple organisms capable of passing through history: bacteria, mosses, dormant tardigrades (arthropods generally found in water), and perhaps, if the seas survive, a few deep-sea creatures difficult to name today (they have barely been discovered). How long have they existed? How do they manage without the energy of the sun? Would the ocean floors provide a haven for creatures suited to high temperatures, sulfides, and everything that we believed to be fatal. Life can be summed up as a capacity for continuous adaptation, a possibility of changing and not persisting in any given state when everything around is changing. What our guide seemed to be saying was, what has to disappear will disappear, and for us that was a direct threat.

Look, he said, these creatures have not known how to evolve, and so they've disappeared!

It's simple. A pause for breath allowed each of us to catch a glimpse of the unrivaled bestiary of the secondary era, the great upheavals of the Jurassic age, of which nothing remains. Layers of sedimentation protect thousands of millions of fossils. We are walking over corpses, doing no more than preparing the next layer to come. Life feeds on life, and in so doing it evolves, invents new forms. Look, if we were to go round the Earth today and make a list of the creatures that inhabit it, a gentleman's library wouldn't be big enough to hold it. Contrary to what most people think, the living world is not fully recorded. Every day new species are discovered. Some appear, others disappear. Our planet remains obscure, complex, full of different ecological niches scattered over the surface of land and sea, but also within the thickness of living matter, from the subsoil to the ceiling of the biosphere. We are living in a magnificent age (his voice rose): never before have the continents been dispersed in such great disorder. It is precisely that disorder that creates diversity.

The guide turned his back, leaving us to stew in silence. We had all involuntarily taken a step back. Not because the lopsided cut of his jacket disturbed us, but because we needed a moment's reflection; how could disorder support diversity? He grew bolder, as he showed us the map of the world:

Looked at objectively, it's chaos, bits dispersed everywhere without reason, north, south, small fragments, large blocks, even almost ridiculously miniscule islands. Imagine all that as a single piece, nice and tidy at the edges, round, harmonious, and you will have some idea of Pangaea, a single sober continent. Today everything is torn apart, in shreds. An amazing tectonic conflict, everything is isolated. That's where the interest lies: geographic isolation creates diversity, because in each of the territories where life emerges it has to invent new forms to adapt to the cold, the dry, the saline, all the different demands of place, all its specific features. "Specificity" through geographic isolation applies at every scale: isolation exists between two continents, between two valleys, between the summit and foot of a mountain, the surface and depth of the seas, the top and bottom of a pot of flowers on the balcony. So, you can imagine . . .

No need for us to imagine. The guide was showing us small examples of endemism, licking his lips as if he were biting into a piece of honeycomb. He gave us a definition of endemism: a constant manifestation dependent on a specific territory. By extension, an individual can be endemic to a continent, a region, sometimes to one valley. Which means that in his natural state he only exists in that part of the world. Everyone is endemic to some place. It is geography that forges uniqueness. For example, I as an individual am endemic to Lavaleix-les-Mines, Haute-Marche, a cool para-oceanic climate. Then he turned toward one of the alcoves:

That thing over there is the ugliest, my favorite. It's found only in the Namibian desert between Moon Plain and the south of Angola. The first botanist to discover its existence uncovered an extraordinary biology. *Welwitschia mirabilis* looks like a mop frayed at the edges, it grows from a double generative stratum and during its whole life—which can be as long as and even exceed two thousand years—produces only two leaves which the wind and passing antelopes reduce to ribbons with brown brittle ends. Its heart, a sort of gritty cork, lies hidden beneath the greenish mass of foliage from which sprout wonderful cones, perched tightly packed in felted scales and studded with moving stamen. An archaic form of fruit typical of the *Ephedra*, a botanical family poised somewhat precariously between the immemorial conifers and our recent angiosperms. In a word, a superb monster. Everywhere in the world diversity is collapsing and being modified. Endemic organisms such as this, confined to a narrow ecological niche, are threatened with extinction. Life wipes out creatures that are slow, those whose biological range remains limited. . . . I'm sorry to tell you, but you are some of the last human beings to observe the *Welwitschia* in vivo. Unless of course people begin frenetically breeding this species, to cover the deserts of the world. But that type of project will always be very unlikely unless some fundamental and trivial reason—such as the dollarization of *ephedra*—were to justify the enterprise, if you follow me?

The well-ordered garden provided alcoves that isolated subjects, like a series of places, distanced from one another in the same way as continents and islands.

Take these tortoises from the Seychelles (they are not dead, they are sleeping). A species with shells patterned with black diamonds set in beige. Ornamental among the rocks but totally boring. I won't hide the fact that I'd be upset if the animal world ended up being represented by nothing other than these inert carapaces. There is a risk. Along with *Arachnida* and beetles, this species has an indecent way of surviving over time. They become animated at meal times and when they mate. They're quite passionate, the act of mounting is accompanied by raucous cries that generally attract the sympathy of spectators. The guide raised his eyes to the sky: How can anything be so clumsy?

The guide bent down to address a few encouraging words to Sidonie—according to him the name of the largest one—letting us see that in spite of everything he was fond of the animal.

Beside the tortoise enclosure there was a cactuslike plant, a remarkable tree from the dry forests of Madagascar. The guide conjured up the thorny savanna of *Alluaudia* (in Madagascan *fansilotora,* octopus trees), home to the enormous *Antenor,* a rare and majestic butterfly from southeast Madagascar. That is all typical, he said, though rare, a whole family—the *Didiereaceae* of which this tree is a member—is found only in Madagascar. Notice how the leaves turn into spines and overlap one another in a spiral, the way in which this plant is sometimes silhouetted against the sky like Aloalo, the totems that decorate the tombs of the Mahafaly tribes; see how this universe distinguishes itself from all the others it might resemble: the semi-deserts, arid tropical regions of the entire world. That too is endemism: a landscape unlike any other, sustainable as long as the preconditions for its existence continue to exist, that is to say, its "insularity" in relation to an increasingly banal and uniform world. At this point the guide stroked an imaginary continent almost despairingly, then made a series of little leaps imitating the *macacos*, comic lemurs from the forests of Tulear, to show us that all was not lost. Finally he made one great leap in the direction of Easter Island, a more distant alcove—loosening the floorboards in so doing, waking up the tortoises, and attracting the attention of a few additional visitors looking for specialist commentaries.

This sweet little *sophora*, admittedly a bit stunted, is a miracle of science and nostalgia. Could it be possible one day to rediscover the beautiful forest that covered the island with this species? (There is evidence of the existence of forestation, whereas today Easter Island is covered with sub-shrubs and grasses brushed by the Pacific winds.) This is the story: a single example, yes, the last survivor of the species (he could have been a street hawker praising the virtues of some domestic object, no doubt endemic too) was able to be saved from the massacre. It was from this one plant that all the others were bred. Thanks to Scandinavian enthusiasm supported by German expertise and French interest, we are now in possession of a sufficient number of fine *Toromiros* (that's their name) to begin the reforestation of Easter Island. An unprecedented operation that the Chileans absolutely refuse to accept, so afraid are they of reinforcing the identity of this tiny territory seeking independence. Forestation, yes; giving the Easter Islanders a flag, absolutely not. Today, the ecological fortune of the planet depends on the human species and elections. That's why we introduce this fragile plant—the sweet little *Toromiro*—as a totally subversive object.

The guide turned to face us accusingly, his eyes hidden in the shadows of his face.

—It's a political act, he whispered.

A sudden fear drew the group closer together. Might it be possible that we have such a responsibility toward the world?

Our plantigrade pulled up the collar of his jacket, ruffled his hair, slipped over to southeastern Australia and declared, pointing at himself: this is what a Black Boy looks like. Standing beside him, the *Xanthorrhoeae* from Perth, with its black body and hair like a firecracker, seemed to have mimicked the guide, forgetting only his stomach: a strange tree, with its cylindrical trunk repeatedly charred by Australian bush fires, a species endemic to this island continent. There followed a long speech about Australia's admirable resistance to time—a continent that he considered to be essentially a fragment of Gondwana. Not the largest, but the one that had remained isolated longest from the rest of the scattered continents and preserved an ancient flora and fauna. To travel in Australia today is to go back thirty million years to the very beginning of a rupture splitting the supercontinent of the Southern Hemisphere into six pieces—Africa, South America, Antarctica, Madagascar, and India. Australia seems to have traveled through history more slowly than any other region of the world—the only one to adopt a marsupial population that was previously shared. Marsupials covered Gondwana. The coming together of the northern and southern continents of America was a threat to the marsupials of the south, less prolific and aggressive than the *Eutheria* of the north. Apart from a few opossums and other small tree dwellers, none of those mammals have survived.

That's what a Black Boy is (he was pointing to the map, to the island pushed away by Tertiary drift), a complete history of the earth. And the history of those who inhabit it. There are maps centered on Australia, with the north at the bottom. When you look at the world upside-down, the map of the sky changes and the gods too.

Geographic isolation favors endemism, that is to say, the originality of creatures, their uniqueness in the universe. It also favors the originality of ideas. For people who have the Southern Cross in the position of the Great Bear as their celestial point of reference, it is normal that their way of seeing the world, and consequently of representing it, assumes different forms.

The Carousel of Myths

He was inviting us to follow him into the universe of man and his beliefs, leaving behind the dragon tree and the *Anona* (a genus of dicotyledons, including the custard-apple and other edible fruits), which we would have liked to know more about. We shall come back to learn more.

Among the creatures isolated by oceans, mountains, and thick forests, we have to include the gods. The gods don't meet except to make war. In front of you are six little pantheons, six faiths, six different reasons for ruling over the world and forcing those who think differently to pay for it. In the middle, a palaver tree, reputed to be the gathering place of ideas and melodies common to these cosmog-

Balinese effigy dedicated to Dewi Sri, goddess of the harvest.

onies. Idealistic, improbable: a species to be identified. After biological endemism—the diversity of species, the carousel of the continents—here we have ideological endemism, the diversity of ideas, the carousel of myths.

The roots of a belief are revealed in its corresponding cosmogony. This is where in each case it is possible to understand and verify the positions accorded to man and nature, in relation to one another. For each culture, their relationship determines the codes of behavior that govern everyday life: we live what we believe. In the long term, that is what threatens (or safeguards, depending on the practice) nature, humans, or both. The diversity of ideas reveals the diversity of the "feeling for nature." What is our feeling for nature today?

Similar in every way to the first garden, the measured tempo of this garden revealed figures and cosmograms enshrined in vegetation. The guide bowed before the figure of Hildegard of Bingen:

There's someone I worship, it's a long time since anyone was so demented. The lady had character and perseverance, she outwitted the popes and the great minds of her day to subject them to her visions, the devil couldn't have done better. An enlightened visionary, at odds with the obscurantism of her day, Hildegard composed music and edited a pharmacopoeia which is still a source of reference today. She denounced the order of the world as it appeared to her in visions. We know of nine of these: each one tackles the relationship of man to the cosmos, the power of God the creator. This one, the third, reminds us of the man inscribed in a circle that Da Vinci was to draw four centuries later. Look at the bright colors, the central position of the son of God with everything else converging on him: the blast of the winds, the force of the elements, the animals, the whole of nature. This selective view of the world was bound to lead to "Manpower," his desire to dominate everything that moves around him. That's the Middle Ages for you, trying to represent life by christianizing the whole universe, right down to the mind of an ant. Fantastic. Each plant was related to the order of good and evil: nature, split in two, was generous with both good and bad. A few centuries were to pass before that came to an end and nature found herself at our feet, distanced, encyclopedic. During that period she still enjoyed an intimate complicity with mankind, the spirit of separation had not yet struck, Copernicus and Newton were not yet born, it wasn't possible to keep at a distance, you had to accept full immersion and submit judgment to the laws of superstition. To assume everyday life to be governed by beliefs and superstitions, there have to be rites and objects, symbols and sacred places; every civilization has produced its own. Everything you see there—the guide was pointing to the cosmographic symbols and ritual objects—all that sacred inventory does more to explain the hole in the ozone and air pollution than any objective analysis of the environment. The way we live our everyday lives is an expression of our convictions. For some people, water is what is used for washing cars, for others it makes the rice grow, for others still it purifies the body. The person who washes cars doesn't think about the rice, and even less about the rituals; the river that flows below his house doesn't have the same meaning for him that it has for the farmer or the priest. It's not only their use of water that separates them, it's an idea of water, a culture of water. The universe is not what we see, but what we believe. Beliefs generate practices. In our country—and increasingly worldwide—drinking water is being sold for the price of wine. We're beginning to worship it like Bacchus. All we have to do is to find a god for it. This present century will provide huge publicity. May I offer you a little water at 14 francs a half-liter, tourist rate, with disposable plastic packaging? Can someone quickly recycle this poly-vinyl for me? No? Perhaps we should change our myth?

A silence, like a bubble, during which someone coughed. Lucien blushed and turned his attention to the imaginary bubble which was supposed to absorb all incongruous sound. Here, the harmony and weight of symbols prevailed. We would soon have to speak in whispers.

To be an ant in the world of the Lady of Bingen or among the aborigines of Australia is a completely different thing. In the first case, the flightless creature risks being burned at the stake for aiding and abetting chthonian powers; in the other, the situation is com-pletely reversed, the creature imposing on man the range of its bio-logical territory, to which we must add whatever oneiric space may be appropriate. In the country where "green ants dream," disturbing the soil is equivalent to destroying the world. Without a doubt, no civilization on earth is as remote from the West as that of the aborigines. In one case, beliefs are accompanied by a design worthy of the convictions: a basilica does the job well. In the other, not a trace, or at most a drawing in the sand. These two examples from opposite sides of the world are enough to demonstrate fundamental differences and oppositions. Is man part of nature? Or is he not? It's almost impossible to compare these two opposing cosmogonies from that point of view. Each time we try, it's as if they were differ-ent worlds. We can do no more than take note of it and be sure of one thing: the universe (one could also say, the "landscape") is what-ever we decide—what we choose it to be. That's why it remains with us, even after we have closed our eyes.

Our guide, his eyes fixed on some distant point, was—we felt sure—contemplating a landscape known only to himself. In short, he was applying the rule suggested by this place: position yourself in front of the icons (there were benches), approach the remote cul-tures through the most radical aspects of their faith—the most fun-damental as well. Even in this profane situation, the ritual objects were unable to confine the myths solely within the dimension of stories and decorations. A vague superstition prevented us from touching them, as if each of them, or each copy, still possessed a magic power. The Ark of the World, a Dogon boat not big enough to hold a newborn baby, the Daughter of the Bison, the Kachina doll of the Hopi Indians, or the medicine staff of the Yanomami shaman, all objects for interceding between the visible and the invisible, offered evidence of the power of a living faith, and all of them in their own way summed up a universe. Perhaps we had stood longest and a little more serenely than elsewhere in front of the great Balinese pre-gembal (an iconic structure created to celebrate the gods at weddings and in temples). That was due to the baroque charm of its construction: a cosmographic image in which the dominating principle is that of the tree and the mountain—in this case Mount Agung—embracing the infinitely complex pantheon of Hinduism-Buddhism practiced in the island. Each object represents the friendly relationship that the people of this region of the world seem to share with the gods: one an offering to the demon of good, another of equal value to the demon of evil.

Don't be deceived, the guide intervened, everything over there is too beautiful (he raised his head, his hands behind his back, like the statue of a military commander). Too much value attached to decoration, too many overly elaborate temples, too much meticulous design, right down to the pillars of the rice barns and the pig troughs. They wouldn't hesitate to make lacework out of the wings of the ibis if that might please the gods. Bali seduces weak spirits, casts a spell over them. Behind all this show, here as anywhere else humans share the common lot. A mixture of gentleness and violence. What other race, so proud, so confident of possessing the truth, so concerned with maintaining harmony with nature, the elements and the gods, would have been so successful in ordering a mass suicide to escape the enemy? I'm talking about a war against the Dutch just a century ago. Apart from the rock of Masada where the Jews perished before Rome, human history is short on sacrifices, preferring a well-ordered confrontation. The Balinese mind under siege deploys a resistance strategy where religion comes to the fore, adorned like a magnificent cuirass in the face of spiritual turmoil. In Bali the original image of the creation of the world is that of a tortoise supporting the universe, encircled by the coiled serpent Antaboga whose breath is the source of life. What could be more solid or intended to endure? This stable and controversial image, said by some to have been reclaimed in order to unify beliefs beyond the disparate religions of Indonesia, commands a prominent place in princely gardens, at the base of sacred buildings, in front of the doors without lintels—an image of the mountain split in two to let the gods pass through. After escaping the evangelism of Muhammad (a moment's silence on the part of the guide to check whether we are listening . . .), here they are, twisting the sacred texts of tourism to invent a way of judiciously exploiting the cultural void. The Balinese lead the crowds from one site to another, and the crowds, delighted by so much kindness, touch upon but never fully grasp such a well-protected culture. In Bali, religiosity pervades everyday life, it is deployed on important and unimportant occasions alike, down to the most insignificant domestic events: crossing a threshold, turning a corner, entering a house. . . . On the day in the Balinese calendar devoted to metal, offerings are made to Toyota and Macintosh. The "surf and sun" travelers are parked in huge five-star boxes built like temples. Meanwhile, the inner precinct of the real temples preserves Shiva, Brahma, Madadewa, and the innumerable throng of deities ruling over humanity, untouched. One should be suspicious of gentle forms of fundamentalism. But in the end—here the guide seemed to relax—these people know what to do, everything is beautiful, even down to their smallest gestures, the rice rituals, the figurines dedicated to Dewi Sri, goddess of the harvest, the art of planting the rice paddy with the first four grains placed according to the cardinal points where the fundamental deities reside. The wind, the rain, the fire, the sun . . . imitating the gestures of a planter, our guide appeared to perform the sign of the cross.

A fleeting hint of insurrection destabilized the group. Some found his speech lacking in respect, others used to traveling insisted that Bali was not at all as people imagined. They could prove it with photographs of that wonderful cremation in the village of Petulu one evening at dusk. We were the only whites, they said, wide-eyed, remembering the excitement of the adventure. It's just as I thought, mumbled the bear.

INTERMINGLING

Wind, currents, animals bring together creatures living at great distances from one another. Man, the essential vector of such meetings, accelerates the natural process of intermingling at the global scale. His action reduces and sometimes removes the barriers of isolation. Intermingling is a threat to diversity even though it produces new situations and new creatures.

Natural Intermingling

So much for diversity. His voice now masked the noise of footsteps and visitors—a combination of sounds particular to the cultural exhibits dominated by the hiss of commentary, and studded with those apologetic gestures that measure the limit or extent of knowledge. On the one hand, he said, biological diversity through geographic isolation is the natural history of nature; on the other hand, diversity of beliefs through cultural isolation is the cultural history of nature. By nature, as you will have realized by now, we mean all living things, man included. But isolation cannot withstand evolution. Not only are the continental plates continuing their navigation and preparing for another encounter (let's say, some million years from now), but also people living in isolated regions always end up at a certain moment leaving their homeland and allowing the winds and ocean currents to carry them to new countries.

Nothing is guaranteed to remain beyond the reach of the flux that stirs up the biosphere like a vast soup; certain seeds travel in

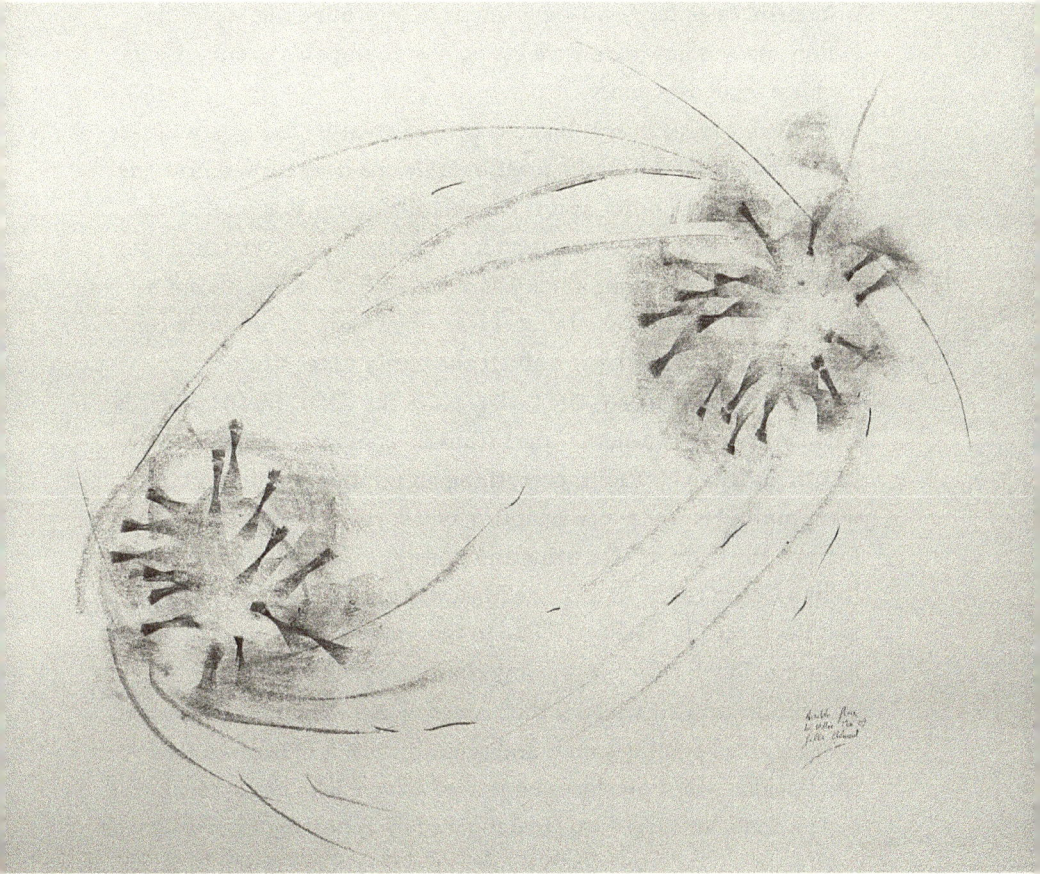

Interchange. *Drawing by Gilles Clément.*

the intestines of birds, so anything is possible. Part of the evolutionary mechanism, and by no means the least important part, the guide assured us, returns to the great global washing-machine, a receptacle into which everything appears to have been thrown in together, its movement continually accelerated by human beings, the unrepentant agitators. In relation to this intermingling, isolation is becoming more and more archaic, the sign of a past condition. In that part of the garden (he was pointing to the first enclosure) we were surrounded by resistant autonomies, frontiers, and identities; now we're moving toward a more contemporary state of the planet, a wonderful cacophony, an experiment in intermingling, a jumble. Follow me, see how inventive life is, how it prepares to surprise us.

First of all, the dodo.

Clucking from a cage. Inside a winged creature as big as a Christmas dinner eyed the world questioningly, his head turned first one way and then the other as if the landscape, being at any one time only half visible, always required its complement. We watched him watching us. In our opinion it was a turkey, but Lucien, always a stickler for detail, insisted that it was a hen turkey. Our guide cut short an obscure discussion about the parity of fat stock.

Here, as you can see, the turkey plays the dodo. These two birds have nothing in common. Apart from this pompous air and a real lack of aptitude for flight, everything about them is different. He was pointing to a life-size sculpture of a sort of obese duck, today extinct: the dodo of Mauritius and Rodriguez, remote islands in the Indian Ocean to the east of the Mascarenes that include Madagascar and Reunion. The turkey, I'd like to remind you, is a native of North America. The distance separating these two pedestrians is in the order of 20,000 kilometers: their meeting impossible. Their connection occurred posthumously and scientifically. A Mauritian tree— the *Davidia*, which produces large seeds, capable of being swallowed by the dodo, let's say—was in danger of disappearing forever: it was not regenerating itself. People suspected that there might be a relationship between the disappearance of the dodo and the reduction in the population of *Davidia*. Like some other species in the world, its seeds have to be broken down by digestive juices in order to germinate. At least, it was possible to verify that even if the plant could regenerate on its own, it benefited greatly from the creature's assistance. The seeds, of a considerable diameter, which used to pass through the stomach of the large duck, today pass through that of the turkey. Too gentle, too easy to roast, the dodo disappeared, having not increased his distance from humans as the latter increased their demands on the bird population of the island. There we have an example of diversity—represented in this case by our tree endemic to Mauritius—being saved *in extremis* by the exogenous American immigrant, a miracle of planetary intermingling. At least that's a positive way of looking at the balance sheet of encounters, the alternative being to consider the turkey as the purveyor of insipid escalopes in the process of insidiously "McDonaldizing" the honest cooking of France. The choice is yours.

But there are other examples: take this "grass from Laos," *Chromolaena odorata*, a eupatorium that's been growing in tropical Africa for the past century. Judged to be invasive, declared a vegetable pest to be eradicated at all costs, expelled from cultivated fields and woodland edges, and now suddenly rehabilitated. It has just been discovered that it has a beneficial effect on the process of reestablishing tropical forest. Its emergence as a pioneer plant between the forest and cultivated land has for the past hundred years contributed to the advance of forestation. It was thought that tropical forests were irretrievably retreating; now they are advancing. So what should we think about intermingling? One could also mention the case of the China rose, *Rosa moschata,* and its role in helping the high-altitude *friches* of central Chile to reach their forest climax. I'd like to remind those of you who may have forgotten, that the botanical species *Rosa* does not exist in the wild in that part of the world.

The bear's eyes were half closed, and a satisfied grimace, spreading from a point on his chin across his whole face, seemed to affect his whole body—right down to his clasped hands, fingers separated, quivering slightly. The guide lingered, he took as an example the island of Reunion, previously known as Bourbon, occupied by humans only four centuries ago in all its unspoiled endemic richness, transformed today into a planetary collection—to the inevitable detriment of its indigenous species. A classic process: occupation of the more or less flat land, cultivation of sugar cane, and the progressive decline of native flora. Introduction of the retinue of species friendly to humans, meaning what they're most familiar with at the time, the usual "ornamental" plants, bougainvilla, *croton, lantana,* or plants grown for spices, medicine, cooking, and so on. Many of them will escape from cultivated land, from gardens, and will form a new alternative landscape. The "false" vine, a Malayan *rubus*, was probably introduced in the hope of extracting a tropical wine from its fruit. Today, hunted down and declared a vegetable pest, the vine covers three quarters of the island.

Intermingling, the irrepressible process of evolution, endangers diversity. Little movement and peaceful isolation produces strong endemism, maximum diversity. A lot of movement and feverish encounters produces weak endemism, minimal diversity. It's an interesting paradox. Our planet today with its multiple continents has more species than at the time of Thetis, the great Pangaea. But it could soon have fewer, intermingling having the same consequences as a realignment of the continents into a single form, virtual in practice but biologically true, a single continent.

Here is one example, this island of Reunion, a summary of the world. From the pounding of the waves on the shore to the top of the mountains, as they say over there, the vegetation is stacked like plates from 0 to 2,500 meters—beginning with the Australian *filao* (*casuarina equisetifolia* or Australian pine), and ending with European gorse. Between these two, the indigenous flora, a crown of relict woodland at about 1,500 meters, takes over from the American and South African levels, while Malayan thorn bushes are taking over everywhere. There is very little left of the colorful woods and heaths of *philippina*, but this indigenous flora, in addition to that of other continents, increases the total number of species found here and anticipates a scenario of intermingling in the process of achieving stability. Right now we are at the beginning of a system of encounters, a process that humans accelerate daily. The final balance sheet should result in a deficit. But in this great adventure, homo sapiens, you and I, take turns with the elements. Look at these anemophilous (wind-pollinated) seeds designed to travel with the wind, or this football of a coconut—the largest seed in the world—destined like the fruit of the Indian almond tree to float across salt seas and take root on the first tropical shore it reaches. The coconut palm didn't wait for humans before invading the world. Within its biological parameters—what is called a biome—a species can become cosmopolitan. No evidence to the contrary.

There was murmuring around us. Some people thought that the guided visit was too guided. Others were pleased with it. The group was getting larger, finding it more and more difficult to push its way through this space congested by intermingling: the garden, laid out in oblique islands, expressed this idea of encounters and energies in tension linking distant points of the planet.

Here and there, a mixture of species from opposite horizons contributed to the composition of the garden as much as to its message: Mexican nasturtiums and New Zealand muehlenbeckia, a landscape found on the edges of Wellington; tibouchina from Uruguay, a reminder of the American layer of Reunion. But most surprisingly in Paris itself: here at the end of the métro line, on any piece of abandoned land, you come across ailanthus and buddleias from China, artemisia from Siberia, acacia from America: an exotic environment long since appropriated for our use as if it were an integral part of our heritage.

That is why, the bear continued, scientists wonder about the best way of preserving biodiversity (in their terms, genetic heritage), that's why radical supporters of indigenous ecology advocate the eradication of imported species (in other words, migrants), that's also why—and I want you to listen carefully to this—the big laboratories record those species likely to generate profitable business (before they disappear), for which they might apply for a patent. The issues of diversity can be summed up as the best way of subjecting the population to the monopoly of a lobby of go-getters. It's an everyday story. If you're looking for guilty parties, just turn round and you'll see a handful of them.

The Traveling Botanists

The other half of the garden, dedicated to intermingling, exhibited personalities and objects connected with travel. Traveling baskets in which rare plants spent their endless journey on board sailing vessels.

An astrolabe, a herbarium, a piece of furniture with trap doors that transformed this chamber of greenery into a cabinet of curiosities. And then the portraits: Hatshepsut, Garcia da Orta, Joseph Banks, and the rest of them. So here are the authors of deliberate intermingling, those responsible for the collections, conservatories, botanical gardens, national herbariums, fishponds, greenhouses, and zoos; the makers of inventories, the great explorers, striving toward an understanding of the world through the collection of its riches, to the point of exhausting the very idea of exoticism. Who today is excited by a rubber tree growing in the airlock of a glass container? Any well-kept town has at least one tropical garden, a rock garden, an aquarium, everything needed to sort out the world and deliver it to the visitor as a package. But before all these curiosities arrived to adorn our daily lives, people were satisfied with travelers' tales and a certain mystery surrounded the exact origins of spices and perfumes. The Arabs, great traders, ensured the trade in rare plants and foodstuffs, taking care to conceal the origin of these riches to preserve their monopoly. Hatshepshut, a patron of distant expeditions, believed that myrrh and incense came from the Land of Punt, a mythical country identified today as either the island of Socotra, south of the Arabian Peninsula, or like Ethiopia, the Horn of Africa. Being inhospitable, these lands could not possibly have accommodated these species sustainably—in reality they came from the Indies and the Sunda islands. Most probably, it was a stop on the route to the Indies. From the time when the powerful queen of Egypt had the plants' journey from Punt to the Nile engraved on a tympanum at the temple of Deir el-Bahari— about 1500 B.C.E.—up to the first Portuguese circumnavigation of the world, followed by the Dutch and the English, the secret of cinnamon, cloves, and nutmeg remained well hidden. Three thousand years of uncertainty, false maps, and naval battles to guarantee those international pirates control of the market and of the seas.

The confrontations over spices that afflicted the great naval powers between the sixteenth and nineteenth centuries can be compared with the conflicts that laboratories engage in nowadays to obtain vaccines. The same issues are at stake. It's just that today they're focused on different speculative resources: petroleum, exotic woods, free cold water fish, whale oil, rhinoceros horn, and so on.

Lucien, the land-locked sailor, dreaming of being rocked by the roaring forties, had stopped in front of a caravel which for him evoked the music of the sea, something that he would never experience, the whistling of the shrouds, the slump of the mainsails in heavy weather, the rasping of the halyards against their moorings, the polished floor of the deck making the cabin boys slide from one rail to the other: a range of words learned by heart, as we knew, and produced at random in the excitement induced by the hope of traveling and the sight of objects intended to fire the imagination. We knew that in the village on a Sunday he would switch the sound system onto heavy seas and invite people to visit his room of models.

To stop the takeover by the lobbies, we would need to set up immediately an ethical committee to intervene against this patenting of the living world; to infiltrate the labs with members of Greenpeace; to place diversity within everyone's reach and destroy the commercial monopolies controlling the living world. Who, I ask you, owns the air? Who owns the water? Madam Pharaoh, he said, addressing the Egyptian, you are playing a game with knowledge that is not without its consequences for the future of the living world, for diversity and what is at stake.

And you too, Mr. de l'Ecluse, he added, intentionally confusing the great patrons of expeditions with the explorers, discoverers, and traders (up to us to sort them out, it wasn't easy). Banks, the naturalist accompanying Cook on his great voyages, would not have traded his discoveries. A gentleman and a botanist, Banks respected science and scholars. Through his intercession, the important collection of La Billardière and the gardener La Haye, assembled during the course of voyages in the Southern Hemisphere on the *Astrolabe* in search of La Pérouse (a French botanist sent on the *Astrolabe* to explore Pacific flora, who disappeared in 1788). The collection was confiscated during battles with the Dutch and finally restored to

France shortly after the Revolution. A significant number of plants bear Banks's name, and a botanic species of the Gondwana family of *protea* is dedicated to him. For Carolus Clusius, things were perhaps different? Europe hit by the Renaissance was exposed to humanist ideas, but the natural world did quite well out of it: tulip bulbs changed hands at extravagant rates, a single plant being worth the price of a house, with the addition of a pair of horses and a few bushels of grain. Never in the commercial history of plants had the market value of a flower reached such heights. Stories are told of the ruin of certain Dutch collectors and the fortunes made by others. *Tulipa clusiana*, its white petals striped with pink, is named in honor of Clusius, to whom we owe the first detailed descriptions of the potato—a plant from the Andes—and of numerous other European species little known at the time, classified by him according to criteria based on their potential use, particularly in medicine.

In the second half of the sixteenth century Portugal launched a number of expeditions, suddenly putting the Far East in contact with South America, with numerous experiments in acclimatization and plant improvement. Garcia da Orta, a physician-gardener living in Goa, then a Portuguese colony, studied the natural riches of the East Indies while practicing medicine. He experimented with different crops, among them the mango, which he improved through grafting. He revealed the origin of cinnamon—the bark of a tree belonging to the laurel family. An analytical appraisal of his work, a little-known book called *Treatise on Herbs*, appeared in 1562. Clusius was to extract numerous pieces of information from it on the nature of medicinal plants; the Latin version appeared in France in 1602 signed by Antoine Colin, who added his own chapter to the first two attributed to "Monsieur Garcie du Jardin" (*orta* meaning garden in Portuguese). Through a series of improvements, the "manga" from the coasts of Malabar in the Indies became the sweet mango that we are familiar with today: oval, fleshy, slightly fibrous, and smelling of turpentine-like floor polish. Orta also introduced the jackfruit, cashew, clove, and ginger—which also traveled via Brazil.

Human curiosity is not restricted to scientists; artists are also interested in nature. When Roberto Burle Marx, as a student at the Beaux-Arts in Rio, drew exotic plants in the greenhouses of Berlin when he traveled there around 1930, and discovered to his amazement the flora of his own country, he immediately wanted to know more and to explore its forests. During the course of his career as a painter and active landscape architect, he was to set up a number of expeditions, revealing to his fellow countrymen the unheard of richness of the Brazilian flora. He was to discover an impressive number of species that bear his name today. Burle Marx used the indigenous flora and transformed it through his artist's eye, the immediate consequence for his country being the appreciation of a material previously despised. He showed how to use plants found within easy reach. But here one detects an exported art, a European influence, an *assemblage* that shows an oblique understanding of nature for what she is, diverse, and of culture for what it demonstrates, a choice within nature—the whole taking the form of a garden designed by a painter.

Quietly, looking at Lucien standing wide-eyed in front of him, the guide said: I wanted to introduce this term. And, then louder: Assemblage! Next space!

ASSEMBLAGE

Man can live anywhere.

Other living creatures cannot.

They regroup in habitats (biotopes) that correspond to their needs.

Each major planetary climatic zone defines a biome, a collection of compatible forms of life.

Within each biome are found a multitude of biotopes— natural assemblages—together with organized territories— cultural assemblages.

—You will have understood by now how intermingling creates the conditions for a new landscape and threatens diversity, increasing the specific richness on a local scale and reducing it globally. Adding and subtracting at the same time. The nature of what has been subtracted is known to us: the total number of creatures living on the planet at this moment is diminishing. The nature of what has been added is less clear. New associations, new ecological features, new marriages, testing the biological range inherent in each species, and so on. What is appearing takes longer to be revealed than what is disappearing. Sometimes that worries us.

The guide looked grave, as if to indicate that in order to have the correct approach to the present it was necessary to be aware of the drama being insidiously played out by the millions of creatures on the planet, without being able to offer any alternative happy scenario. Our group drew closer together, united in their concern for everything that was going wrong on earth. A child passed by with a "hyper" yo-yo, sending out sparks and strident music. He was showing us the way.

The last space in this sequence was laid out as a landscape: water plants, trees, grasses.

A rice paddy in the center, a peat bog on the edge, two environments totally different in geometry and appearance, united only by the presence of water. A deliberate choice to illustrate two very distinct methods of assembly. One, artificial, widespread throughout the planet, organizing a specific landscape; the other, natural, rare, only a few examples worldwide.

—This is a "total garden," said the guide, in this space everything is assembled, but not randomly. Spontaneously, living things produced through global intermingling regroup according to their biological affinities, their compatibility in terms of living. A plant originating from tropical Africa will only survive in America or Asia in tropical conditions. Natural regrouping always takes place within what is called a biome, a region of compatible living conditions. The tropical biome of the rainforests is found on the planet in the area

between the tropics of Cancer and Capricorn. In theory, if all the continents were merged, species originating from this biome are capable of cohabiting. If they're assembled in a diagram like the one you see here, you get an image of the biome concerned in its entirety. It's easy to appreciate its importance in relation to other biomes recorded on the planet. You can see, for example, that the Mediterranean biome occupies a limited territory in comparison with the tropical biome or that of the hot arid deserts.

The guide was right. Even by adding to the Mediterranean biome a piece of central Chile, a tip of South Africa, a fragment of western Australia, a few scraps of Tasmania extremely thinly dispersed in the mass of California—all regions considered to be Mediterranean—

you got only a tiny cornerstone, seemingly forgotten among the huge stack of deserts, northern forests, and other tundra. The whole thing looked like a badly trimmed top that a child wouldn't want to play with, but the adults seemed to find it amusing. We gazed, fascinated, at this virtual continent on which we were supposed to live from now on. This image troubled us. It was not simply the result of scientific research but a new way of seeing the world. A single continent, this was what our territory was, an indisputable (biological) continuity between environments of similar affinities, no matter how great the distance separating them, or the immensity of the oceans. If that was the case, if the living world was really organized in this way, we would have to utterly rethink our cartography, our way of explaining the world, its development, its economy, its politics. It appeared to turn everything upside down. From now on we could consider all those countries sharing the same climatic zone as a single biologically compatible unit. Instead of referring to the continent of the United States where there is a mixture of cold and hot and humid or arid zones, we would have to speak of biocompatible "United States": half of Brazil with most of Gabon, the whole of New Guinea to which we would have to add a scattering of islands and a few Asiatic peninsulas, and there you have the humid tropical biome. This game of assembly can be played with all the other biomes, to see whether, having regrouped the disparate countries in this way, there might be the embryo of a common cultural element, a way of living the climate, for example. Faced with the living requirements suggested here by the strength of the biomes, the cultural frontiers of mankind seem to depend on minor circumstances and an arbitrary tangle of regulations. One wonders how they have withstood the test of time so well. The other thing that surprised us was that this reality seemed to have been discovered only now, in this century, whereas it felt as if it went back to the dawn of time, as if things had always been this way. Continental drift, plate tectonics, nothing more than jolts in a tense mechanism, consistently uniting, finally quite simple. The earth being merely a kettle at its core, the

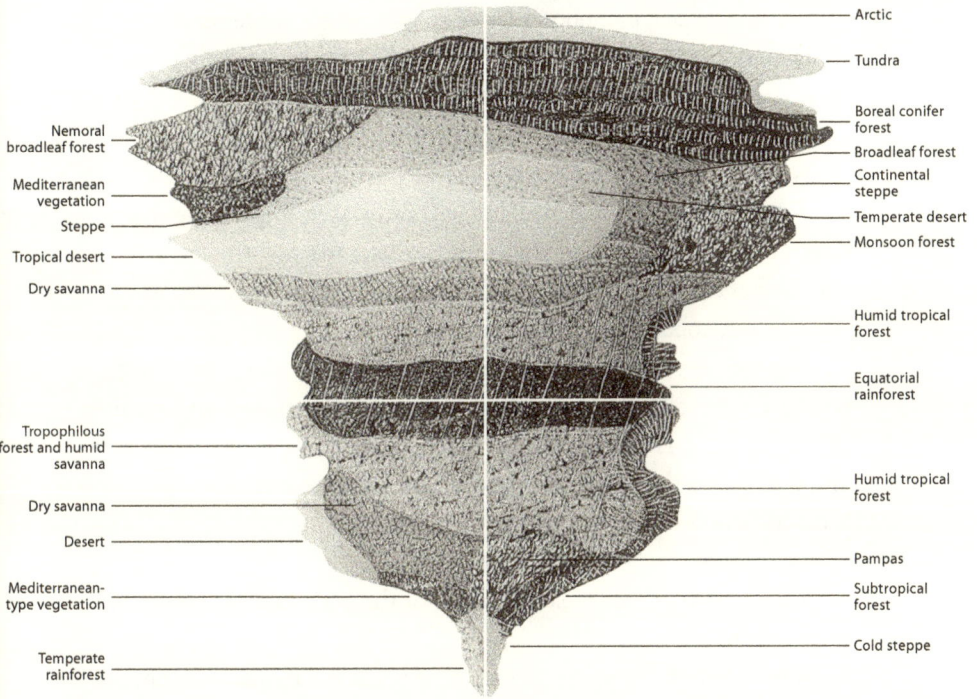

Theoretical continent. Superimposition of the major climatic zones with their assumed vegetation, irrespective of continents. Virtual Pangaea corresponding to a possible biological reality as the result of planetary intermingling.

continents floating lumps, sometimes bound together in a single mass, sometimes scattered on the surface of the magma. The whole thing burning continuously since the beginning, and for a long time to come. Some people were touching themselves as if they needed to find a possible correspondence between the human body and this image of the continent. The head was there alright (although a bit flattened), but the feet crammed into the tight slippers of the southern continental mass made the whole edifice look as if it were resting on points: the designer must have forgotten the Antarctic. That was true, what could possibly correspond to the glacial biome, home to Emperor penguins and Kerguelen cabbages?

—And the sea, somebody dared to ask, where is the sea?

Lucien, frowning, went up to the guide; we were standing in a circle, he was pointing to the icon bereft of oceans.

—There is no such thing as land without sea.

We all agreed. Lucien knew what he was talking about, a planetary garden without beaches, life without water, impossible. We began to have doubts, had the guide perhaps reached the limits of his competence and pushed the discussion beyond what was admissible, purely to assert his authority?

Impassive, planted ever more firmly on his back legs, ignoring our bewilderment, the bear was cutting a piece of paper with a pair of nail scissors. He folded the form in a concertina, wide on one side and narrow on the other—it was a small virtual continent—placed it on the water of the rice paddy and observed:

—It floats.

We gazed at the unbroken surface of the little aquatic field transformed into an ocean.

A member of the maintenance staff retrieved two mobile phones and an Instamatic from the bottom of the pools, objects clearly outside their biome. The garden compared two methods of assemblage: in the foreground the orderly organization of the rice paddy; further away the apparent confusion of the marshland. Would the territory of the planet in the future be entirely dependent on the principle of "management" on the one hand, or "reserve" on the other? In one case, humans assembled according to their criteria for maximum profit, in the other nature assembled with the aim of deploying maximum diversity. Planetary gardening, if there was such a thing, should offer alternatives. Was it possible to imagine a territory that was natural, diverse, and profitable? The bear showed us marvelous photographs of landscapes where nature organized by humans produced "gardenlike conditions" of almost infinite variety. We were reassured. In the two examples chosen, the process of intermingling was redesigning the landscape. In the case of rice, an Asian cereal, one could equally evoke an image of America or the Camargue. In

the peat bog, *saracenia*, a carnivorous plant from America, was growing alongside *sundews* (*droseras*), reflecting the existing situation in
the high-altitude peat bogs of Europe.

—The world is changing, the guide insisted, but for living things
the slightest change is matched by requirements. Things can't go
anywhere or anyhow. Rice can't yet be grown high in the Alps. He
bent down to blow on his paper boat and concluded:

—It's not profitable.

A single species of rice—*Oryza sativa*—covers the world and
feeds it. It can be seen growing on industrial plains like wheat in
Beauce, or "gardened" as on the terraces of Asia. In the first case
the "agricultural park" denies the site, at most complying with the
land register, the violated landscape becoming more artificial. In
the second, the "agricultural garden" does no more than emphasize
an existing condition, considered indispensable because it is natural. Landscapes that have been anthropomorphized, that is to say
controlled and fashioned by humans, can usually be recognized by
a simple strict geometry—ponds, reservoirs, terraces—whereas
nature is organized in a complex and seemingly barely legible way.
They reveal a social organization and furthermore, a way of seeing
the world, of surviving in it. But it is possible for the human presence to be effaced without actually ceasing to interact with nature.
Certain types of African forestation, certain Amazonian forests that
appear to be wild, prove to be gardens. Places where there is control
and respect for life.

—Assemblage operates according to the unchanging laws of
evolution: possible or impossible. For all nonhuman living creatures it depends on the biological range of each species and com-
poses biomes whose limits coincide with those of climates. For
the human species, the only one to be exempt from the rules of
distribution that govern living things, it is expressed by a certain
"naturalization" of the landscape, that allows management politics
and cultural choices to confront one another. One cannot compare
the endless expanse of a crop of genetically engineered corn on an

American plain with a pocket-sized rice paddy in Bali. Humans are everywhere, they take with them the plants and animals that they find useful, accelerate intermingling, and acting as if for themselves with a view to increasing resistance to the harshest environmental conditions, they modify the genetic makeup of the living things in their possession, orient the living world. They garden. The current state of the planet is the result of these three processes: endemism, intermingling, assemblage. But the profile that our landscape will finally assume depends on the way Man regards Nature, the position he occupies within it, the hope that he invests in his territory; it depends on gardening.

Crouched over the rice paddy, Lucien was blowing the paper boat in the other direction, the game being to make it cross the pool between the stems of rice. Seen like that, curled up in a ball and folded, our little craft seemed small. Vulnerable. It was obvious, a breath could make it change direction.

Meanwhile, the bear had disappeared, leaving us in front of the "Garden of Experiments," as if we had suddenly become responsible for ourselves, real gardeners: nine trestle tables loaded with objects to be handled, screens and texts, summed up thematically the nature of the experiments. Observe in order to act, that's what determines our experiments and makes them acceptable. The planetary garden cannot be subjected to traditional cartography. It is everywhere, it occupies the biosphere, its territory comprises the the multiple layers of living matter.

Leibnitz's great idea is that . . . nothing in the world can be separated.

—GILLES DELEUZE

Nature: a physical reality existing independently of man.

—LAROUSSE DICTIONARY 1988

The limits of life on earth, the biosphere,
these are the limits of the garden.
Man, omnipresent, responsible for all living things,
he is the gardener.
The layers of the living world, earth, air, water,
that is the territory
At the heart of the garden, the uncontrolled forces of life
*   and its inventions,*
the dream of man and his utopias,
both defining from one day to the next the unpredictable
*   trajectory of evolution.*

—GILLES CLÉMENT

We have never stopped not knowing what nature is. What is
different today is that we are dealing with a planetary nature.

—ISABELLE STENGERS

When one knows nothing about nature, when everything appears to be incomprehensible and scientific, accessible only to specialists, what does *political gardening* mean, when even the tiny plot where the vegetable garden tries to express itself raises a thousand questions and leaves as many answers hanging in the air?

Shouldn't we begin by considering the extent of the territory, fixing the limits of our gardening? Then from within a defined enclosure, try to understand before intervening, pause long enough to assess the parameters of the place, its complexity, and finally how to be part of it? How can humans inhabit the earth without condemning it? Two approaches to dealing with living things without causing them harm:

—Observe in order to act: an analysis whereby all future action comes down to gardening, as we understand the term here. That means that the operational phase—the intervention—is not an ideological response dreamt up to manage or save the planet, but on the contrary, relates to the *specific case,* seen as a local ecosystem whose model is never exportable.

—Work "with" whenever possible, "against" as little as possible: a way of thinking that allows all gardening to be organized with a view to the greatest economy of means. It presupposes accepting and sometimes even developing ways of collaborating with energies already present, principally those that nature offers in every circumstance and region of the world. It implies being sparing in the consumption of negative energy and if possible doing without it altogether.

"Observing in order to act" and "working with" provide the foundations for planetary gardening, whatever the scale of the enterprise. Each person can adopt his or her own method and develop it according to his or her capacities and needs. It calls for common sense and forces the imagination to consider what is most biologically sustainable: a project, in which life is interpreted as a continual reexamination of the present.

As to the extent of the territory, its limits, we have to refer to the layers of living matter, to its components: air, water, earth. . . . One cannot map the planetary garden, it doesn't coincide with any recog-

The dirigible that directs Francis Hallé's treetop raft floating above the tropical forests of Gabon, 1999.

nized physical or political boundaries. Nevertheless it is possible to imagine the limitations to human actions on the territory. We know, for example, that the greenhouse effect can be reduced by regulating emissions of carbon dioxide from the burning of fossil fuels. On the other hand, the great natural forces of the planet, such as El Niño, are among the phenomena over which humans have no control. The limits to gardening are fixed by the limits to the extent of man's power over nature.

Starting from that basis, it is possible to determine the actions that allow humans to organize their territory while attending to living things. That means, in the long run, attending to humanity.

But it would be unwarranted and dangerous to envisage organizing this territory by subjecting every nook and cranny to human "gardening." To preserve its flexibility, its reserve of inventions and possible reversals, in a word its *future,* the garden must preserve an area of uncertainty, a space of nonintervention sufficiently diverse and extensive to allow it to regenerate as soon as the need becomes apparent. Nature spontaneously finds alternative solutions to the constant reorientations of evolution. We can be confident in her powers of invention. For that reason above all others, we must allow her a real share in existence.

Which leads us to a precautionary principle: how are we to succeed in preserving the source of all energies while we ceaselessly draw on them?

Is it possible to take without impoverishing, to consume without degrading, to produce without exhausting, to live without destroying? The practice of gardening responds to these questions, precisely by observing a precautionary strategy. At harvest time the gardener will not lift and consume the whole crop; he will be careful to put aside a portion destined to produce viable seed for future crops. He will never allow the soil to become exhausted, erosion to destroy his land, or water to be poisoned.

Do actions exist at the global scale comparable to those that the gardener adopts in his garden? Can one transfer the vocabulary of the garden, usually associated with spaces that are restricted and enclosed, to a space seemingly immense and open?

If the planet functions as a single living entity, limited by the confines of the biosphere, then we do indeed find ourselves in the conditions of a garden: an autonomous and fragile enclosure where every factor interacts with the whole and the whole with each of the creatures present. All that remains is to find the gardeners.

When Francis Hallé, botanical explorer and director of the treetop raft project, proposed to exploit the tropical forest without ever destroying a single plant, he invented a form of gardening on a grand scale that responded to two principles:

not to abstain from the exploitation of nature's riches; and
to preserve those riches in their entirety.

This ambitious project involves an intimate knowledge of the
environment. Taking turns on the "raft" are scientists in charge of
uncovering the complex functioning of the forest, and others whose
objective is to discover active molecules useful to medicine, cosmetics, and so on. The collection, harvested with the help of a "treetop
sled" suspended from an airship, enables them to count the species
inhabiting the canopy many of which are still unknown. The equipment invented by the architect Gilles Ebersolt—the raft, the sled,
the icosahedron (a nest in the trees)—make it possible to enter the
environment without ever damaging it. To give an example from the
gardening perspective, the plan is to remove trees without opening
up any roads—from the sky—each tree taken out being replaced
with a new planting. A futuristic project, anticipating the day when
sooner or later this method will have to be used to exploit the tropical rainforest, at present seriously endangered worldwide.

When mayor Jaime Lerner decided to organize the sorting and
recycling of waste for Curitiba, capital of the state of Parana in
southeast Brazil, on the basis of barter, he involved the entire population in an act of gardening that consisted of

reducing the production of waste; and
trying always to recycle it.

The slogan of this town of more than 2 million inhabitants—
O lixo que nao e lixo (waste that is not waste)—transformed a simple
daily gesture on the part of its inhabitants into a political act, each
of them participating, whether consciously or not, in a truly global
citizenship. The whole of the town—its urbanism, transport, educa-
tion, and immigration policies—responding in an exemplary way to
the issues of this century, which are all to do with ecology.

From our point of view, Francis Hallé and Jaime Lerner can be
considered as planetary gardeners, one planning to garden the earth
without destroying it, the other to organize the gardener's house

so as to offer the best possible conditions for man and nature. Both have chosen a positive option for the future: to save the whole body of living creatures and their territory. According to the analysis offered by geneticist and philosopher Albert Jacquard, humanity, for the first time in its existence, is in a position to decide its future: to live or to perish, to ensure its future or to commit suicide. Curiously, the decision eludes most politicians, too busy satisfying the wishes of their electorate; it also eludes the global lobbies, interested exclusively in submitting the world to their monopolies and making the greatest possible profit—all of them acting in the short term, paying little attention to the consequences of their actions. The most powerful nation announces cynically, with scant regard for the entire human race, that it has the means to buy the right to pollute, wherever it may choose. The right to kill.

The decision comes down to the individual citizen. To be hostage or actor, there is barely a choice. If considering intervening, he or she has to decide on a method and a philosophy. In terms of a way of thinking and a management strategy, the planetary garden offers a certain number of possible actions. Some of these are already happening on the planet. But there are boxes full of research awaiting an executive decision before it can be carried out. At a time when the monster nuclear energy provider EDF finds itself face to face with European competition, the diversification of energy sources is becoming a reality and, so to speak, obligatory. When a large French town is scared by a project for the autonomous management of an urban park—calling solar pylons artistic madness—it is not only taking a step backward but covertly fighting against the planetary garden (and yet we are talking about a *garden*). It reflects in a caricature manner the archaism of institutions and the inability of public authorities to anticipate the future. But when the "gardener of the clouds" stretches the nets in which fog condenses to provide water for a village in the Atacama desert in the north of Chile, he is doing no more than taking inspiration from the rain tree on the island of Hierro in the Canaries, experimenting in one of the numerous areas

Hallé's manned treetop raft in the canopy of the Gabonese forest, 1999.

opened up by bionics—the science that mimics nature. By inviting the inhabitants of his village to use water fairly and with respect, he is making individuals responsible for their territory; he is gardening.

The concept of "working with" opens up unexpected opportunities and changes the way we look at our environment. In the Mascarene Islands there is a "cyclonic wood" where flowering is triggered by leaf-stripping storms: it is the cataclysm that ensures the survival of the species. Elsewhere, midwinter frosts release from dormancy certain species that without this vernalization would disappear. In many regions of the globe, particularly in those with a Mediterranean climate, fire decimates and regenerates the landscape. Its physicochemical action in releasing seeds from dormancy allows all these pyrophytes (literally: vegetables of fire) to regenerate. Too long a period without fire causes a senescence which in the long term can destroy them.

So clouds, wind, cold, and fire are all assets that the gardener can interpret as tools for planetary gardening. Very quickly it becomes apparent that the sum total of factors defining a biocenosis (or ecological community) in a given place must be considered as allies. An observer, free from the "green as a golf course" neurosis where his pretty lawn is concerned, will accept the burrowing of the mole that helps to germinate species which, without this opening up of the soil, would remain forever as invisible seeds.

All the known experiments in planetary gardening cannot be grouped together under a single heading. It is possible to establish a typology of projects under way, referring constantly to the fact that the gardener invokes the sky, but does not control it. It is always a question of working in collaboration with nature.

This task can be declined in nine tenses:

1. Not harming the earth
2. Welcoming the gardener's allies
3. Promoting the exchange between living creatures
4. Knowing how to manage water
5. Building the human house
6. Preserving the gardener's enclosure
7. Caring for the earth
8. Giving nature its share
9. Producing without exhausting.

The sections that follow have been written with the help of the collaborators on the original Planetary Garden exhibition.

NOT HARMING THE EARTH

Human beings have always known how to enrich the soil in order to improve its productivity, notably by adding nitrogen in a form easily absorbed by plants. However, the race for productivity has led to the massive use of fertilizers that have exhausted the soil, damaging the natural mechanisms that ensured its regeneration. In certain newly productive countries, land clearance has resulted in fragile soils unable to support the intensive agricultural methods of countries of the temperate Northern Hemisphere.

The population of the planet is growing. It is no longer possible to depend on traditional farming techniques, particularly in developing countries. At the same time, it would be dangerous to impose the productivity model regardless of the cost. Another way must be found to put in place techniques that satisfy the needs of humankind without condemning the soil. Such techniques do exist: they often rely on traditional methods that had been set aside in favor of spreading fertilizers. A new front has emerged, aiming to develop an agricultural economy respectful of the environment and capable of ensuring sustainable growth.

From Satellite to Fly Trap

The *glossina*, or tsetse fly, a carrier of parasites causing devastating effects on humans (sleeping sickness) and cattle (loss of weight, drop in milk production, abortions, mortality, incapacity to work, etc.) infests a large part of sub-Saharan Africa. The resulting losses, estimated at around $1.5 million a year, considerably reduce agricultural development, the supply of protein, and consequently the economic health of numerous poor countries. For a long time, the only measures available were the clearing of vegetation and local deforestation, combined with the control of wild animals by slaughtering. Then, insecticide sprays became the most common method (applied on the ground or by plane or helicopter), used over vast surfaces and with a certain success. But the current concern to apply cleaner and less expensive techniques tends to favor trapping methods (mechanical traps, "living traps" in the form of cattle coated with insecticides) that interrupt the transmission cycle of the parasites by considerably reducing the density of stinging insects.

The CIRAD (Centre of International Co-operation for Research in Agronomic Development) has set up a project in western and central Africa, aiming to identify environments favored by the tsetse and

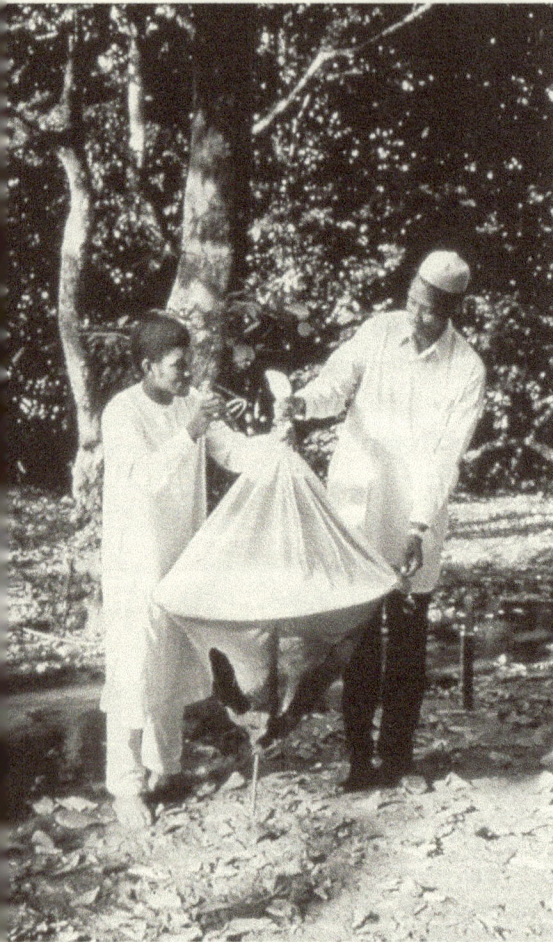

Fula cattle herders of Burkina-Faso setting up a biconical fly trap impregnated with insecticide, made by Challier-Laveissière.

frequented by cattle, thereby enabling the control to be targeted at those priority places or interfacial zones. The "targeting" is achieved by matching data from ground surveys (natural vegetation, agriculture, soil, water, livestock, producers, etc.) with data produced by analyzing images from the Satellite for the Observation of the Earth (SPOT). The rapid diagnosis of situations at risk opens up the possibility of local rather than general management of the trapping systems by the farmers whose rise in living standard is often dependent on cattle.

Direct Sowing:
Sowing Without Plowing

In the past thirty years the savannas of the humid tropical zone of Brazil, the *cerrados* of the west and midwest, have seen a process of transformation comparable to that of the prairies of the American Midwest during the twentieth century. A form of intensive agriculture has been set up, usually a monoculture. But the reclaimed soils are poor and acid. If they are plowed with mechanical equipment, they deteriorate, erosion wreaks havoc, agricultural returns diminish. To avoid this waste, new methods of cultivation have been proposed. The first thing is to protect the soils from erosion with a permanent covering of vegetation, either living or dried. At the beginning of the rainy season millet seed (for example) is scattered. When it has nearly reached maturity it is scorched with herbicide. Soybeans are immediately planted, thanks to a technique that makes it possible to sow directly through the dried out blanket of vegetation. Once the soybeans have been harvested, millet or sorghum is planted on the residue following the same method. This is harvested, and the straw flattened underfoot, ready for the dry season. The quality of the soil is ensured through strong, deep rooting. Different crops are planted alternately during the same year, but may also be replaced by pasture for longer rotation periods. This means a shift from a monoculture to a diversified type of rural landscape, infinitely less aggressive toward the soil and respectful of biological and climatic balances. These techniques are currently being adapted for small-scale farming in Madagascar and in certain African countries. This approach is suitable for use in other more temperate regions confronting the problem of soil damage. Perhaps one day we'll be eating Roquefort from sheep fed on sorghum . . .

Plants that Fix Oxygen from the Air

With the aid of their roots, or through the process of photosynthesis, plants extract from the soil or the air everything they need to live (hydrogen, carbon, oxygen, nitrogen, sulfur, phosphorous, sodium, potassium, iron, etc.). Nitrogen is an element that plays a fundamental role in the chemistry of living things. But plants, like humans, are organisms with no chemical mechanism to enable them to directly absorb nitrogen, an essential component of the air. To satisfy their nitrogen requirements, plants have to acquire nitrogen in an easily absorbed form, that is to say, in a mineral (ammonia, nitrate) or organic (manure) form that their roots extract from the soil.

However, monocellular organisms exist that can break down the molecule of atmospheric nitrogen into ammonia. They therefore form ideal companions for plants. Symbiotic relationships can develop, particularly in the case of legumes that host and nourish nitrogen-fixing bacteria, *rhizobia*, in their roots and stems. The cultivation of legumes is therefore a way of enriching the soil with nitrogen: this type of "green fertilizer" has been known since antiquity.

In order to reduce the use of chemical nitrogen fertilizers (expensive and polluting), research has been undertaken to maximize the use of these natural fertilisers, such as *Sesbania rostrata* or even *Azolla*, a small fern used in Vietnam since the eleventh century to fertilize the rice paddies.

WELCOMING THE GARDENER'S ALLIES

The gardener and the plant are not lovers isolated in the landscape. The earth that sustains them is teeming with life. Knowledge of the multiplicity of living things makes it possible to garden in harmony with nature anywhere on the planet. The plant plunges its roots into the earth and sways in the wind. A spade thrust into the soil reveals a population of worms, millipedes, spiders, grubs, and an invisible mass of microorganisms, fungi, and bacteria.

On the surface of the soil this fauna moves about frenetically in search of food. By breaking down the organic matter, they make it easier for the roots of plants to absorb. Moreover, the earthworms contribute to the aeration of the soil by digging their tunnels in it. As for moles, real policemen of the subsoil, they devour the harmful grubs.

Humans also make fire an ally. In different regions of the world, fire is used to clear the ground before introducing crops. South African horticulturists know the influence it has on the regeneration of the *Fynbos* (the scrubland of the Cape). They use its heat and smoke to germinate the seeds of numerous ornamental species.

The tree canopies of the tropical forests, like the depths of the ocean, are the naturalists' last terra incognita. Bathyscapes and tree-top rafts will need much more time to develop, at the bottom of the oceans and in the air, to identify the hundreds of animal and vegetable species that are still unknown to us.

Worms to Fertilize the Earth

The worm, by disturbing the ground, modifies the mechanism and structure of the soil. The tunnels that it digs help to aerate it, to facilitate the circulation of oxygen and water. In order to grow, roots can borrow the channels provided by these tunnels, which promote the plants' growth. The worm, a true chemical engineer, also introduces modifications into the chemistry of the soil because its droppings are rich in easily assimilated nitrogen and phosphorous: by burying the residue of tea bush clippings along with worms, it is possible to increase the production of plantations by a factor of 1.5.

Worms begin to be effective at a density of thirty grams of live weight per square meter. It is useful to have a population composed of several species since each one has its specific role in the different strata of the pedosphere (the outermost layer of the earth, composed of soil).

There are about four thousand species of earthworm known today: as they reproduce by parthenogenesis, it takes only a single individual to generate a whole population. A lot remains to be discovered, particularly in the tropical regions.

The Garden of Fire

In certain regions of the world fire has been appearing naturally for thousands of years. Consequently, numerous vegetable species have adapted to it. These species, known as pyrophytes, have developed different biological strategies enabling them to survive.

Adaptations to fire are sometimes so advanced as to become indispensable to their regeneration. In South Africa, when a fire has not occurred in a given area for a period of thirty or forty years, certain species become senescent and disappear. Fire maintains the open spaces they need and also induces the germination of their seeds. In effect, the heat and the smoke it produces act as physico-chemical signals that break the dormancy of the seeds. For many years South African growers have made a habit of treating the seeds of pyrophytic species before sowing them: soaking them for a moment in water at sixty degrees Centigrade, heating them in a frying pan or smoking them in a hermetically sealed container. In Australia, the southeastern United States, and South Africa, managers of natural parks intentionally start controlled fires in order to regenerate certain ecosystems.

Because fire has a detrimental effect on the well-being of modern man, by destroying his surroundings and material possessions, it is perceived as a catastrophe. A less anthropocentric approach to this natural phenomenon prompts us to reconsider it: incontestably, in certain ecosystems it maintains landscape diversity and biodiversity.

The Treetop Raft

The tropical forest is the reservoir of an abundant biological diversity. Certain elements (vegetation, animals, insects) are a potential source of molecules that will form part of tomorrow's medicines, phytosanitary products (that is to say, they protect crops), and perfumes. A large part of this biological activity takes place in the trees where the light is brightest. The treetop raft makes it possible to explore the canopy, which remains one of the most unrecognized zones on our planet. Consisting of an enormous platform perched on the top of the trees with the help of an airship, the "raft" enables scientists to set themselves up in the forest canopy to carry out their research. Their exceptional biodiversity makes tropical forests a priceless resource. But if the destruction of the forest continues at its present rate, we will soon have lost a large part of this treasure without ever having had the opportunity to discover it. In order to limit and contain deforestation, new economic interests have to be created, so that it is more profitable to preserve the forest than to destroy it. This is the aim of the work being undertaken by Francis Hallé, leader of the treetop raft missions. Eventually, the forest should bring in more through the economic development of its biodiversity than from the exploitation of its wood.

PROMOTING EXCHANGE BETWEEN LIVING CREATURES

For some time now, ecological chains, notably the relationship between prey and predator, have been scrutinized with all the modern scientific equipment available. The chemical and genetic war waged by plants and insects for the past 250 million years is beginning to reveal its secrets. From now on, that knowledge can be exploited to increase revenues by reducing the losses caused by predators, without recourse to industrial pesticides.

The gardener is going to encourage the reproduction and development of these biological agents. They will provide a helping hand in limiting the proliferation of pests. It is a matter of exploiting nature without doing any harm, by drawing inspiration from traditional practices and the most recent biological knowledge. Careful observation of natural habitats suggests ways of acting, often simple ways that consist of exploiting the antagonisms and alliances between organisms.

The Acadja

The *acadja* is a traditional fishing technique used in the coastal lagoons of West Africa. This very productive way of fishing can enable a catch of ten to twenty tons of fish per hectare per year. The *acadja* promotes the concentration and breeding of fish by introducing bundles of branches that form an artificial reef that encourages the development of epibionts (seaweed, periphyton, zooplankton), the natural food of fish. However, if allowed to develop in an uncontrolled way, the practice of *acadja* can have disastrous effects on the environment: excessive forest clearance, erosion of banks, organic pollution of the habitat through the decomposition of the branches. On the Ivory Coast and in the forests of Guinea, the Institute of Research for Development is studying ways of exploiting the benefits of the *acadja*, while avoiding its negative impact on the environment. In order to limit deforestation, the use of bamboo stems seems to offer a promising future. Given that bamboos regenerate

Gathering fish caught in the acadja, *Guinea.*

much faster than trees, this solution would be a way of saving the forests. Furthermore, an *acadja* made with bamboo produces a return comparable to that of the bundles of branches, with the advantage of a life span four times longer than that of the wood. The absence of bamboo can be remedied by setting up bamboo plantations near the *acadja* sites. An extension of this promising experiment is currently being developed in Guinea. This example could be used in other regions sharing the same geographic and socioeconomic context.

The Hedge, the Pear Tree, and the Parasite

Very often, a plant has a regular predator, an insect or fungus that feeds on the chemical vegetable products specific to it, or takes advantage of its physiological structure to find shelter. This can result in the plant being severely damaged by the presence of the parasite. The pear tree *psylla* is an example of this type of perverse bilateral relationship: its larvae cover the branches, their secretions scorch the leaves and cause fungi to develop that damage the fruit. Numbers of insects known as entomophagous (insect-eating) feed on this *psylla* and can therefore be used as a means of biological warfare. Generally however, orchards are introduced into an environment very low in useful insects: they are often managed as a monoculture or surrounded by other chemically treated crops.

So the biological fight against the pear tree *psylla* involves creating a varied environment of vegetation in the vicinity of the orchards, likely to attract a rich biodiversity of useful insects, particularly specific predators. The best solution is a hedge made up of a mixture of various broad-leaved species: ash, willow, laurel, Judas tree. The pear tree *psyllas* are also preyed on by insects with a more varied diet, such as ladybirds. So one can plant hazel, elder and lime trees, and ivy, which will provide active allies against aphids and the pear tree *psyllas*. The spread of flowering periods according to species is another way of ensuring both constant food for the predatory insects, shelter during the cold season, and a stable population.

KNOWING HOW TO MANAGE WATER

Water is indispensable. Without water the soil is infertile. Without water life is impossible; it is the medium within which the chemistry of life unfolds. Water is also a terrestrial and atmospheric element, unevenly distributed across the surface of the planet. In one place, long walks through the desert for a few liters; in another, an orgy of baths, watering, flushing.

Considered in rich countries as an inexhaustible and cheap resource, water contributes, through intensive spraying and the massive use of fertilizers, to the growth of the vegetable biomass of highly profitable crops. To the point where, in those places where there is no shortage it becomes a resource polluted by soil contamination. Yet elsewhere, water drowns, floods, ruins, brings sickness and death through the parasites that it harbors. In still other places, its scarcity is cruelly felt.

The relationship between humans and water has consequently been, and always will be, ambiguous. In order to be harmonious, there has to be a point of balance difficult to define, an exercise of judgment that is becoming a priority almost everywhere in the world.

The Village of Fog

It takes 30,000 liters of water to manufacture an electronic chip, 400,000 to manufacture a car.

Chugungo is a little village in the Atacama desert in the north of Chile, situated on the edge of the Pacific Ocean and at the foot of the *cordillera* of the Andes. There is no river or spring, and it never rains. For a long time the inhabitants were supplied by water tankers, until the day when a scientist had the idea of transforming the thick clouds of fog coming off the sea into water.

Water nets were laid out on the side of the mountains. So when the wind blows, the fog is pushed through these nets, which retain about 25 percent of its humidity. Little droplets form that are then collected in pipes. It is possible to produce up to 60,000 litres of water a day.

The Salmon of the Rhine

Since the Middle Ages, the Rhine had been a river where salmon were fished in abundance, to such an extent that a law forbade the Swiss bourgeoisie from serving it to their servants more than twice a week. There were still salmon to be found in 1930, but the population decreased rapidly, and finally became extinct toward the end of the 1940s. The fact was that this migratory fish could not survive its passage through one of the most industrialized and polluted regions in Europe. Moreover, man-made adjustments to its bed (canalization, large dams) deprived the Rhine of a large part of its natural fauna. In 1986, the fire at the Sandoz pharmaceutical factory, near Basel, caused nothing less than an ecological catastrophe. The five countries through which the river flows made the decision to clean up the Rhine. A policy of shared action (purification plants, industrial decontamination) was set in motion for this entire region of more than 50 million inhabitants.

In 1991 the countries sharing the Rhine basin approved a program for reintroducing salmon, made possible by the decontamination of the water. As a result, in Alsace since 1994 it has been possible to see that the salmon have returned to the water courses, beginning their characteristic life cycle there once again: born in the small rivers, then leaving for the Atlantic for one or two years, before returning to spawn in their river of origin.

BUILDING THE HUMAN HOUSE

The advance of the western model can be seen everywhere, particularly in developing countries whose populations are rapidly increasing. Agricultural work chains people to the land, whereas towns offer the promise of pleasure and the hope of achieving prosperity one day. Meanwhile, poverty increases.

So action is required on two fronts. On the one hand, the condition of the poor megalopolis has to be improved. That presupposes the existence of a municipal authority capable of taking initiatives and able to count on the collaboration of its citizens. On the other hand, limits have to be set on rich towns' reckless squandering of raw materials which increases pollution.

Humanity demands the "right to a town." This "right" must be conceived and formulated in a spirit of planetary solidarity. Seen from this perspective, the four towns selected here illustrate different ecological approaches: Kalundborg and its industrial ecology; Stockholm and its ecological tradition; Curitiba and its social ecology; Porto Alegre and its political ecology.

Kalundborg

Population of the planet in 2050: 10 billion.

Kalundborg in Denmark is the most successful example of an enclosed industrial system. On a single site, five industrial companies exchange their waste products, in such a way that the waste from one becomes the raw material of the others. It includes an oil refinery, a power station, a pharmaceutical laboratory, a chemical products factory, and a plasterboard factory.

This system functions according to the same principle as natural ecosystems: nothing is lost, everything is transformed. At the beginning twenty-five years ago, two businesses were involved in this exchange. Today there are nineteen different flows of exchange: nine programs of water recycling; six programs of energy exchange, and six programs of waste recycling.

The program for water recycling is typical. There are very few subterranean water sources in the region: so as not to exhaust the groundwater, the Kalundborg industries exploit the waters of Lake Tisso. This water is used by the oil refinery for cooling, before being redirected to the power station to produce the energy required by the refinery and the chemical products factory. However, the power station uses coal, an operation producing emissions that contain sulfur.

Embarking on a new form of exchange, the station has selected a desulfurization process whereby the waste—gypsum—can be used by a neighboring industry. With this gypsum, plasterboard is manufactured. Even if this association, based on commercial agreements, was not originally conceived with an ecological aim in mind, it nevertheless demonstrates today in an exemplary way that it is possible to protect the environment on an economic basis.

Stockholm

Because of their special relationship with nature, the Swedes developed very early on, at the first signs of the effects of industrialization, a very strong ecological awareness. Stockholm, an archipelago city, has been concerned with the problem of water since the 1930s, and has subsequently practiced an aggressive environmental policy based on close cooperation between universities, research laboratories, and economic players. That is why businesses, pushed by strict consumer demand, are careful to use "clean" manufacturing processes and the least polluting products. The town policy, to give it its rightful name, is characterized by a unique cooperation between the municipal companies in charge of energy production, water purification, and waste disposal, the aim being that they function as an "ecocycle."

So the district heating network that supplies 200,000 homes—the largest in the world—gets its heat supply from units of recycled water, the residual sludge being used for the production of biogas. Air and water have always been the primary concerns of the city. The development of shared transport systems, the prevalence of buses running on ethanol, the restriction of harmful emissions through the common district heating system, ensure exceptional air quality for an urban environment.

Water, treated according to very sophisticated processes, is currently cleansed of 50 percent of its nitrogen and 95 percent of its phosphorous before being returned to the sea.

Stockholm is one of the only capitals where you can swim in the center of town, and where restaurants serve fish caught in local waters.

A completely new residential quarter is under construction in the heart of Stockholm on a former industrial wasteland: Hammarby Sjöstad. It has been planned by the local authorities as a model ecological town.

Curitiba

The population of Curitiba, capital of the state of Parana in southeast Brazil, has grown from 300,000 in 1950 to 2,100,000 in 1990. To address this demographic explosion, the municipality has undertaken an innovative development programme, led by Jaime Lerner, former mayor and subsequently state governor. For him, putting into practice a policy that combines economic development and environmental protection with satisfying social needs requires the support and participation of all citizens. The efforts of the municipality were focused on four priority sectors: urban transport, with the introduction of a very efficient bus network aimed at decentralizing economic activity (the proportion of the population heading for the center of town dropped from 92 percent in 1973 to 54 percent

At the troco verde *(green exchange) in Curitiba, Brazil, the municipality provides fruits and vegetables in exchange for recycled waste.*

in 1999); quality of life and the development of green spaces (taking public areas alone, 52 square meters are available for each citizen); employment; and the recycling of waste.

This last issue has been developed in an exemplary way in the *Troco verde* (green exchange). Since the garbage trucks are unable to circulate in the shantytowns, the municipality of Curitiba has set up a totally innovative system for the collection of selected household waste: in return for a kilogram of recyclable waste, the 35,000 families taking part in the operation receive a kilogram of food. In this way, about 250 tons of recyclable waste are collected every month. This program, combining the protection of the environment with social concerns, improves hygiene in the shantytowns and brings a nutritional supplement to an underprivileged population.

Porto Alegre

Porto Alegre is the capital of the state of Rio Grande Sul, the south-ernmost state of Brazil, on the borders of Uruguay and Argentina: 1.3 million inhabitants in a greater metropolitan area that includes more than 3.3 million. The city is run according to a particularly innovative democratic principle: a participatory budget that involves a unique experiment in direct democracy. The city is divided into sixteen sectors with each appointing two representatives. These take their place in the Participatory Budget Council, alongside various represen-tatives of the population and ten councilors assigned to five named committees. This grand popular council organizes plenary sessions that bring together never less than 500, sometimes even 1,000 people. The Participatory Budget Council enjoys total autonomy in terms of deciding, within the budgetary parameters established by the municipality, anything that relates to investment. Setting up a new service or any commitment to an investment depends entirely on the participatory budget, which corresponds to about 20 percent of the overall municipal budget. This system has been introduced so that the population may participate directly in the running of the city and make its concerns known: it is they who, in a totally autonomous way, prioritize its requests for works and services. In various areas—housing, transport, sanitation—there has been a complete reversal of the priorities usually found in Brazil. The participatory budget is not unique to Porto Alegre; this type of management exists in other municipalities governed by the Workers' Party.

PRESERVING THE GARDENER'S ENCLOSURE

The principal source of energy on earth is the sun. It functions like an extraordinary thermonuclear reactor, emitting energy through the splitting of atoms. This radioactive mechanism is the source of the internal heat of our planet.

For millions of years, the light given off by the sun has enabled plants to carry out photosynthesis. It is this biochemical operation that manufactures the living tissues and the carbon elements (sugars and lipids) that feed them. Coal, oil, and gas are the result of past accumulation of this organic matter in lake or marine sediments. These fossil energies are not inexhaustible. Humanity must immediately take steps to find alternative sources of energy. Will the chemists of tomorrow discover how to use the chlorophyll that transforms light into chemical energy, to provide our megalopolises with electricity? One way or another, plants will always remain a renewable source of energy for us. The development of hydrocarbon from vegetable crops has already been mastered (as in the case of sugar cane in Brazil) or is in the process of being mastered (as in the cultivation of euphorbia in desert regions). Meanwhile, solar heat, river water, geothermal heat, and wind remain nonpolluting and underexploited sources of energy.

Water, Wind, and Other Renewable Forms of Energy

Eighty-one percent of world consumption today is dependent on combustible fossil fuels (oil, 40 percent; coal, 24 percent; gas, 17 percent), their combustion emitting 20 billion tonnes of carbon dioxide into the atmosphere every year.

Fortunately, today we are witnessing the development of alternative sources of energy. By around 2025 half the consumption of the developing countries could be provided by renewable energy, used in a modern way.

By doing this, the countries of the Southern Hemisphere would only increase their greenhouse gas emissions by 45 percent, for a growth in energy consumption of 200 percent. In India, for example, wind energy will produce nearly 3,000 megawatts of electricity by 2010, which means the equivalent of three nuclear power stations. In 1998 alone, wind energy installed worldwide grew by 25 percent compared with the previous year (32 percent in Europe and 15 percent in North America). Six offshore farms are functioning today off

Sweden, the Netherlands, and Denmark—the latter estimating the possibility of installing more than 4,000 megawatts of wind energy.

Finally, plants exist that contain hydrocarbons as a natural part of their vegetable substance. *Cobaifera langsdorfi* (also known as the diesel tree) is a large tree that grows in the Brazilian forest. A hole drilled in its trunk will produce a liquid fuel suitable for use in a diesel engine. A hectare of these trees could produce the equivalent of fifty barrels of oil a year.

Naturally, there is not much chance of acclimatizing this tropical species to the Beauce, or introducing it into the Forest of Fontainebleau, but in Guyana—with an area of 8 million hectares, equivalent to one sixth of France—these trees could produce 400 million barrels a year, which is far from negligible, considering that French consumption is in the order of 900 million barrels.

CARING FOR THE EARTH

Nature, tidied up and cultivated, in contrast to the supposed purity of virgin spaces. Man's relationship with the earth is sentimental, economic, possessive, jealous. Today we are a little more aware of the damage that human actions have inflicted on nature. How is this damage to be repaired? The problem is complex and urgent, because urbanization is increasing and the population growing.

How to amend, how to anticipate? The treatment of contaminated soils can be undertaken in various ways that involve botany, microbiology, chemistry, and electrochemistry. One technique is phytoremediation. In order to clear excess metals from the ground, it is possible to grow plants predisposed to absorb them, allowing them to be recycled. Other "vegetal reparation" practices are used, for example, to slow down the advance of the desert. These methods, being subject to the slow rhythm of plant life, require patience and continuous application. This involves long-term investment and civic actions undertaken on behalf of the future inhabitants of the planet. To repair and care for the earth are our duties from now on:

we have to try and leave behind us clean soils, particularly while we still know how and why they have been contaminated.

A Green Barrier Against the Desert

On the coastal rim of Senegal, the sand dunes are advancing by eight meters a year, transforming these regions into desert zones.

In order to stop this advance, the forestry authorities have since 1948 been organizing the mass planting of a tree imported from Australia, the filao (*Casuarina equisetifolia,* also known as she-oak or ironwood). The sandy soil, deficient in nitrogen, would have prevented this reforestation were it not for a bacteria applied to the filaos, enabling them to synthethize nitrogen from the air. In the tree nurseries at Boro, the filao seedlings are sprayed with water to which these bacteria have been added, as a result of research undertaken by the Institute of Research for Development.

The young plants are then transplanted to the coastal dunes. Today, between Dakar and Saint-Louis du Sénégal the green barrier extends over nearly two hundred kilometers. The filaos halt sandstorms and stabilize the soil with their roots, enabling the farmers to grow food crops within the shelter of this barrier.

Decontaminating Plants

Woodland, agricultural, urban, and industrial soils are, in ascending order, the receptacles of numerous pollutants, the accumulation of which can cause problems for food safety and the dysfunction of ecosystems. Among the different polluting compounds that result from human activities, trace elements (copper, cadmium, lead, zinc, etc.) occupy a particular place: unlike organic compounds, they persist and accumulate in the environment.

In certain cases it may prove necessary to treat polluted soils so as to reduce the risk of transferring toxic elements to living organisms, particularly to humans. Metals not being biodegradable, the problem of managing them is difficult to resolve.

Several solutions have already been tried out, using microorganisms and plants. First, phytostabilization, which consists of growing tolerant plants that ensure that pollution is immobilized, thereby limiting the risks of metals being carried down to the groundwater. Second, phytoextraction, a low-cost technology which respects the biological properties of the soil, depending on the cultivation of hyperaccumulator plants that extract the metals or radionuclides from the polluted soils. This scenario would involve harvesting the aerial parts of the plants (above ground, i.e., leaves and stems), drying them, and then incinerating them, after which the ashes could be dumped or reused in metallurgy. The first experiments with phytoextraction are recent. The species generally tested belong to the genuses *alyssum, thlaspi,* and *brassica* from temperate zones.

In the majority of vegetable species, the elements removed are confined to the roots (90 percent), only a fraction being transferred into the aerial parts. Nevertheless there are species that can absorb and accumulate from ten to five hundred times more trace elements in their aerial parts. These hyperaccumulators that can colonize sectors where other species disappear (metal-bearing environments) have for several years been of particular interest to researchers, in the hope of increasing the yields from extraction.

GIVING NATURE HER SHARE

Increasingly, humans live in towns. Nature is admitted there only as an exception: urban parks for walking, scraps of woodland beside roads, tree-lined avenues, flowerbeds at the center of roundabouts. Alongside this official enclosed nature, there is a clandestine nature to be found: ruderal plants springing out of the pavement, taking over a piece of wasteland, spilling over a ruin. Even more enclosed is private nature: the bonsai growing indoors or the pots of flowers on windowsills. "Nature" is a sterilized "elsewhere." How can the citizen be made to understand the value of this "elsewhere"?

The space of the landscape, natural space, is essential to the survival of concentrations of human beings. Their well-being and their equilibrium depend on its being cared for and protected. The forests and countryside surrounding the town have to be managed. So too, those spaces where rain falls. Maintaining a cover of vegetation prevents a too rapid runoff, allowing water to feed freshwater springs. The example of water makes it clear that the landscape, green space, is essential to the survival of large concentrations of people.

The public authorities have decided to define areas of green space for conservation. While fighting against its depreciation, we must also oppose the excessive love that leads to a site being trampled underfoot, with millions of tourists carrying away the fragile earth on the soles of their shoes. In order to give nature her share, to begin to respect her, we must first learn to know her.

The Raz Headland

The site of the Raz headland (the westernmost point of Brittany) was degraded through an enormous tourist presence. The natural heritage, a characteristic heath with fragile stands of vegetation protecting the site from the onslaught of the sea, was subjected to a growing number of tramping feet. With the disappearance of these heaths, the flora and fauna were threatened (for example, the Dartford warbler, the rare red-beaked chough). Erosion was advancing, resulting in the earth being less able to resist the forces of the sea. Certain areas were stripped bare down to the bedrock. Within the framework of Pointe du Raz being named a "Grand Site de France," experiments in reestablishing vegetation were carried out. Access routes and buildings were removed, the commercial complex was torn down, a new complex was built incorporating the landscape which restored forgotten views to the site: the impact on the regeneration of the landscape is considerable.

Barely two years after the start of work, the results are impressive, even though the regeneration of the vegetation cover is only in its early stages.

Urban dereliction.

The Forest of Abandoned Plots

Délaissés is the word used to describe plots of land that have been managed by man during periods of urban development, before being abandoned. What should be done with these abandoned urban plots? A workshop set up by the Caisse des Dépots et Consignations (the national French savings and banking institution that manages local funds) under the theme "a forest for the future" proposed to plant trees according to a proven and inexpensive method: the technique of forestry. In this way, a plot of land that was costing the community money can become profitable economically (through the exploitation of the forest, and the removal of property tax), and above all socially (through the transformation of waste ground into open space). A summary count of these abandoned plots in the Paris region, in Marseille, and in Toulouse produced evidence of the diversity and above all, the ubiquity of these plots. A map drawn up by the Institute of Planning and Urbanism of the Ile-de-France region, indicating the total number of vacant plots, underlined the way they complement the overall network of green spaces. One might imagine an Office of Abandoned Spaces whose mission it was to count and observe them, to set up hypotheses concerning their growth together with a management plan. A voluntary policy might be based on a restructuring, or a more vigilant observation of the forestry codes. For example, article 52-2 of the forestry code authorizes the prefects to define the priority perimeters differentiating " areas to be maintained as woodland." An abandoned urban plot could acquire this status, either by a simple declaration on the part of the owner or by the automatic registration of a perceived fact, through a statement backed either by an institution or by a local community.

The size of each of these pieces is of little consequence: the system of vegetation that develops on a small isolated area of abandoned ground is part of the world of the forest through the network that it forms with neighboring plots.

PRODUCING WITHOUT EXHAUSTING

Chemical pollution by fertilizers and pesticides, dramas about mad cow disease or dioxins, uncertainty concerning genetically modified organisms—modern agriculture raises question marks in the minds of more and more people, provoking anxieties or criticism.

Meanwhile, the farmers, their professional groups, and agronomic researchers set up new practices aiming to reconcile the imperatives of economic competition, feeding populations, product quality, and respect for the environment.

Several achievements illustrate this ambition, notably in the cereal, sugar, protein-rich plants, wine-making, market gardening, and horticultural industries.

Productivity and Respect for Nature

"Precision agriculture" enables treatments to be adapted to the strict needs of the plant, particularly reducing water consumption or chemical products, as well as the pollution that results from them. "Biological warfare" against pests is being developed as an alternative to "chemical warfare." It consists of replacing traditional plant care products with the natural animal or vegetable predators of the pathogenic agents of the cultivated plant (the vine, for example). These useful predators are often already present in the crop or in its immediate environment. Developing oil-producing and protein-rich plants (rapeseed, sunflower, peas, etc.) as part of crop rotation, ensures a better control of nitrogen, trapping the excess nitrates in the soil before they are transferred into the groundwater.

Cogeneration (combined heat and power, using heat transferred from another industry) aims to economize on greenhouse heating and reduce polluting emissions. In certain regions, crops such as vines are used to maintain and even improve landscapes threatened by neglect and degradation. These new approaches and techniques are still in the experimental stage, but are indicative of a significant and irreversible evolution. To take one example: 33 percent of the cooperatives in France's Centre region already have research projects under way—experiments in the field of precision agriculture.

EPILOGUE

The planetary garden considers globalization in the light of the
diversity of living creatures and practices, but is radically opposed to
the standardization of living creatures and practices.

It seeks to integrate the biological, political, and social factors
that interact on the planet, in the belief that a single model of organization is never possible.

It claims for the territory a plurality of ideas and actions, just as
it recognizes an infinite number of ways of gardening integrating
the complexity of life.

It involves the human species as much as the individual, endlessly reminding them of their responsibilities toward the garden.

It proposes a relationship between man and nature in which the
preferred actor—in this case the gardener, citizen of the planet—
acts locally on behalf of and in awareness of the planet.

Life, Constantly Inventive

REFLECTIONS OF A HUMANIST ECOLOGIST

l'Amour Symbiotique
la vallée mai 2009
Gilles Clément

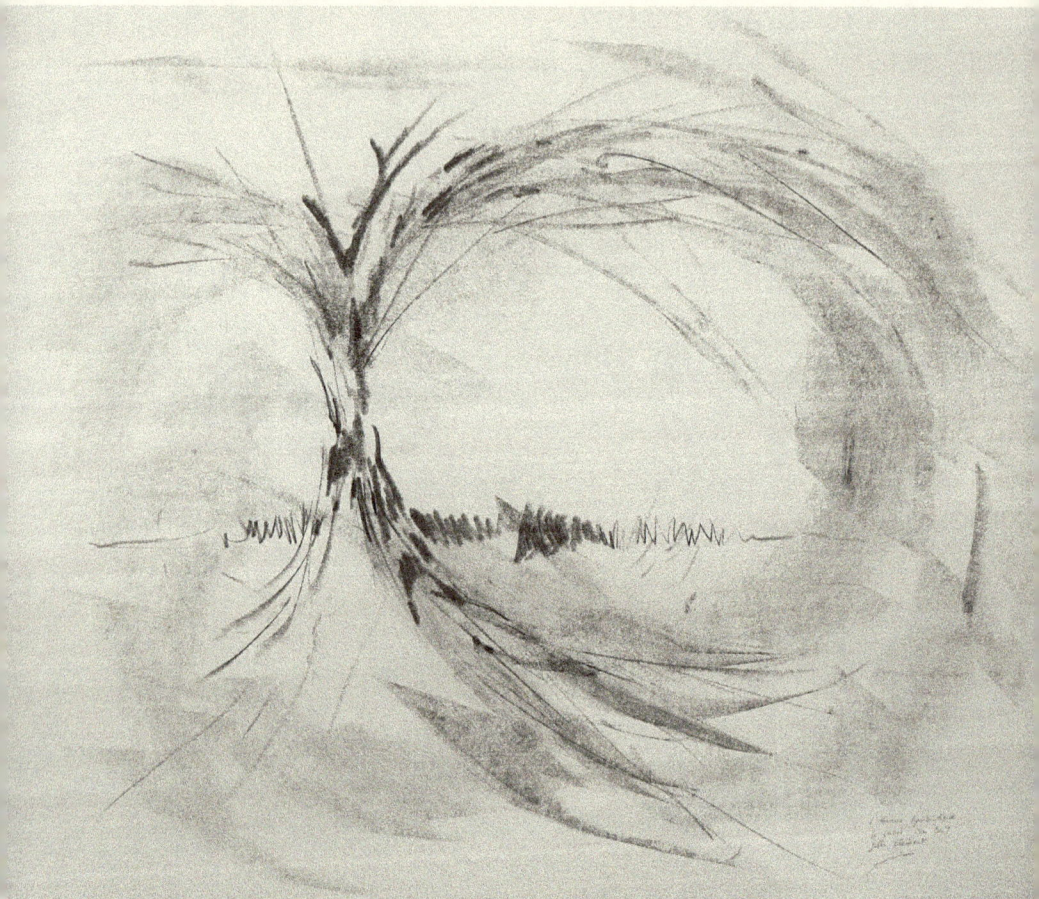

Symbiosis. *Drawing by Gilles Clément.*

I AM A GARDENER

I am first and foremost a gardener: on the one hand, because I have a garden, and on the other, because I think that the garden is at the forefront of our current understanding of the terrain as a whole, and consequently of the landscapes that creep into the garden. I am a gardener in the literal sense of the word when I handle the soil and again, at a certain remove, when I attempt to work on landscapes that are large-scale, but the questions uniting the two are the same: they both revolve around living organisms, and that is a strong point in common that cannot be side-stepped, a question that cannot be avoided.

The real problem is one of demography. There are a great many of us. That determines a system that needs consumers. There must not only be a lot of people to consume, but each one of the consumers must consume more and more. We are always hearing that such and such a country has a population growth that will ensure its economic welfare: madness, even from a system that has decided that in a finite world infinite development would be possible, something evidently impossible. An American economist—interestingly, he is both an economist and an American—says that to imagine this, meaning this infinite development in a finite world, you have to be either mad or an economist. We are currently embroiled in this system that, apparently, nobody wants to put a stop to because it is convenient.

THE WORK OF THE GARDENER IN A SEA OF CONCRETE

Where, in this context, are nature and diversity, important to us since we are dependent on them, faced as we are with this pressure of population, its activity, and the consequently traumatic effect of that activity on the terrain? If we pay close attention to everything that surrounds us, it becomes clear that there is a possibility that this assemblage of nonhuman beings, which are nevertheless our

fellow travelers in our stay on the planet, may be capable of living
with us provided that we treat them sympathetically.

That implies that we do not traumatize those beings that are the
plants and animals surrounding us. It is possible, even in a dense
city, in a very built-up, well-provided landscape, to have diversity if
we are able to understand how it functions and how we can encourage it. Such a vision involves suppressing all those products that
kill—pesticides for example, but others as well. With a simple over-
dose of fertilizer, it is very easy to kill; overdosing is common with
agricultural and horticultural treatments, and that wipes out all
sorts of species, known to be abstemious, that cannot tolerate such
overfeeding and die. – This allows nitrophilous plants to grow, which
specifically need excessive doses of those things, often common
greedy plants, such as brambles, nettles, and docks.

A SCANDALOUS LAW AGAINST NONAUTHORIZED PRODUCTS

One question seems to have been insufficiently discussed, not
having been given the recognition it deserves; this has allowed
that iniquitous law, said to be agriculturally oriented, that was
voted for in January 2006[1] and the legislation endorsed during
the summer of 2006.[2] For me, this law involves the confiscation of
common property. It prohibits the wholesale of products that have
not been authorized (treatments, fertilizers, pesticides etc.). What

[1] This refers to article 70 of the LOA (Loi d'Orientation Agricole), which separates the
risk assessments linked to inputs into plant material, which is entrusted to the AFSSA
(Agence Française de Sécurité Sanitaire des Aliments—a public body created 1999
following the outbreak of mad cow disease) from the decision to authorize putting these
products on the market (Autorisation de Mise sur le Marché or AMM), which remains the
responsibility of the minister of agriculture. This article particularly draws attention to the
fact that, being a question of plant pharmacological products containing one or more
active substances intended for the treatment of plants, "any commercial advertising and
any recommendation . . . can only be applied to products having marketing approval."

[2] This law was amended in December 2006, but in June 2007 the legislation required for
the amendment had not yet appeared.

Making puree of nettles as a natural fertilizer, Jardin d'Orties, Melle, 2007.

is a nonauthorized product? It is a product that does not have the approval of the lobbies—which is to say, it is effectively outside the market. Therefore this is a law made by or for the lobbies. It is forbidden to experiment with, sell, or distribute, even for free or even speak about all those substances that anyone can make at home and for nothing. Today, you risk a fine of 75,000 euros and two years in prison by speaking about nettle manure, a well-known natural fertilizer and weed killer. But nettle manure is only the visible tip of an iceberg, a symbol for everything related to common property that is susceptible to being commercialized. It involves a whole series of extremely important operations that we may well carry out every day, or vital actions that we do without even noticing; in the end, the very act of breathing oxygen will be taxed because this common property will have been authorized, there will be private owners of fresh air and they will keep us under a dome of polluted air (that's easy enough) in order more readily to promote the sale of breathable air nearby. One day we will be sold the right to breathe.

Nowadays, common weeds are prohibited. You can no longer use those nettles for your own purposes, nor speak about them, sell them, give them away, or advertise them in any way. Now that's what I mean by common property. The Americans are in the process of imagining how one might share the clouds, meaning water. Therein lies a fantastic economic potential. Who owns that passing cloud? This is in no way anodyne. True, we are living in an ultraliberal society that allows this sort of manipulation—with the result that the very small businesses that sell these products die, killed off by the large companies. That a state should make a law to endorse this kind of tendency is intolerable, unacceptable; it is something that should rouse public opinion. But opinion is hidden behind a cloak of deceit. On the pretext that it's just a matter of nettle manure, it's nothing very serious, almost derisory. But when the day comes that we have to buy fresh air—and it's the same thing—people will begin to realize that things have gone too far, but it will be too late. We must not allow this law to be passed! Quite simply, we must demand its repeal.

Obviously, I cannot make a petition on my own, but I entirely support the idea of the petition that already exists. I don't think that at the present time the plant protection service is going to come and check whether one is doing something legal or illegal. But, *a priori*, someone who cultivates a plant at home is breaking the law, in the case of cannabis, for example. And the nettle can become just as illegal, since the whole thing is arbitrary.

In the case of nettle manure, it is not just a question of commercialization, but of publicity, meaning, simply making it known that there is a way of treating one's land with free products, very easy to procure and produce, that present no danger and have been proved effective over centuries—you simply have to know the methods and dilutions. To confiscate their use is an attack on civil liberties.

As far as seeds are concerned, they have long been liable to confiscation. It is no longer legal to market plants that have not been recognized as grown from seeds authorized by the major seedsmen. Hence the difficulties experienced by KoKopelli, a small associate distributor of seeds that are not available on the official market, distributing seeds of ancient varieties or varieties that have always existed and proved their worth. There are also new ones that are not yet on the market and never will be, simply because they don't represent a sufficiently interesting return for those who make the money. KoKopelli is in financial difficulties as a result of legal proceedings, because obviously, given the power of the lobbies, it is easy to attack this type of company. KoKopelli is being progressively destroyed by one legal case after another which—whether it loses or wins—cost a lot of money.

A third example, of a similar kind, concerns herbs: there are people who collect plants with medicinal or other benefits and sell them. So they too are prosecuted. A syndicate, SIMPLES, set up in 1982 in the Cevennes, represents eighty producers/collectors of medicinal, aromatic, culinary, cosmetic, and pigment plants, which shows that more and more people are involved in the productive, economic marketing of herbs, alternative seeds, and biological agriculture.

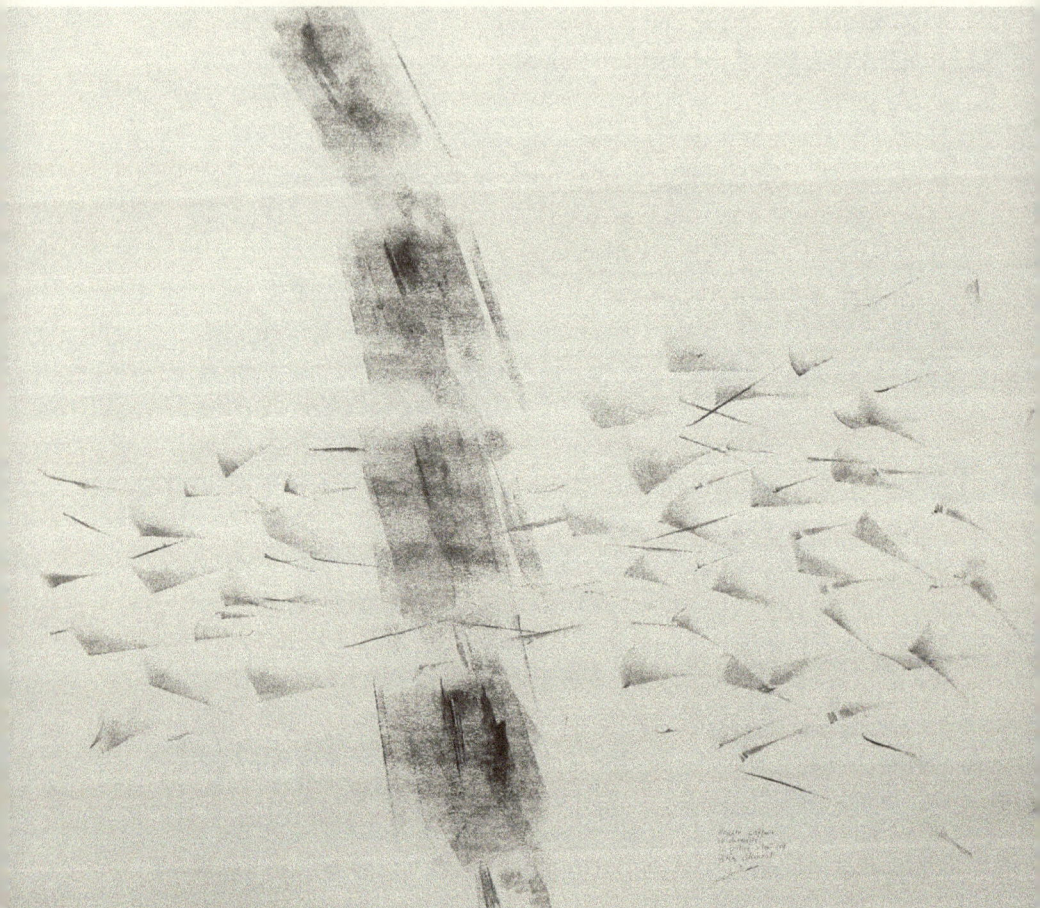

Collapse *(Effondrement)*.
Drawing by Gilles Clément.

That is precisely what the lobbies fear: loss of profits due to the fact that so many people are turning to methods that sidestep them. They cannot create an economy from it. How can a law, said to be agricultural in its orientation, have been passed when it sets out to prevent people from living better on a daily basis even in a domestic sphere, a way of life that could be said to be in harmony with what is pompously called sustainable development, which if it means anything, must be precisely this kind of domestic activity.

COMMUNITIES GET INVOLVED

When it comes to planning, the communes generally have too much money: when they begin to take up the granite paving-stones as they are doing in my area in order to lay traffic humps, it is because there is too much money, and they don't know what to do with the grants. But there are hopeful signs. As the result of a study promoted by the town of Rennes, the municipality has given instructions for forms of direct action to try and get the population to accept pavements not being treated with weed killer. Brittany is one of the regions of France where for some time it has no longer been possible to drink the water, because it has become unfit to drink—whether from a tap or a well. So you have to buy it in bottles. Here you have the polluters on one side and the salesmen on the other (sometimes they're one and the same). That is perhaps what is going to happen with air, with everything. That is why those laws, voted for in the middle of the night, during the month of August when everyone has other thoughts on their mind, must be pinpointed and seriously analyzed, not treated lightly. So the town of Rennes is taking an educational step with the population, to make them accept areas of diversity where nature is free. The aim is to avoid all kinds of treatment, including, incidentally, weeding with flame guns. This example shows that large towns can become aware and put in place really concrete measures.

Each one of us is responsible for the planet. So we need to know how to organize our everyday life; this type of method is profitable in terms of sustainable development. But it is opposed. Why? Because it doesn't bring in money for the large companies. The society in which we live is *against sustainable development*. It's full of these words and does nothing toward it. In fact, quite the opposite. It is inadmissible that a law such as this should be voted for under that name.

The lobbies in question—in my opinion, the authors and initiators of this iniquitous law—hope to produce the chemical equivalents that they will then baptize as "biological" while adding their own label. These will be authorized and will be marketed by them. At the same time, they will sell other substances recommended for killing the nettles that you are not allowed to cultivate, because otherwise you could make your product for free.

A HUMANIST ECOLOGY?

The term "humanist ecology" applies to a way of understanding the relationships between living beings according to the precepts of ecology, without ever excluding humans.[3] It is opposed to the radical ecology, according to which life on earth can exist without humans, which is indisputable, but who would there be to appreciate radical ecology if humans actually disappeared from the planet?

The book that we published recently was initially proposed by Louisa Jones, well known in the gardening world as a journalist and writer. Among those journalists interested in the garden, she is a little different from the others in that she carries out her research on gardeners and landscape architects in depth, with a great respect for their work and ideas. The title chosen for the book comes from the exhibition "The Planetary Garden" held in 1999–2000 at La Villette, which had as a subtitle "A Political Project for a Humanist Ecology."

That proposition gives a hint of what I was trying to evoke, to discover: how we humans can live with the prevailing diversity, and how that diversity, for its part, can live with us? I am presupposing that we are part of nature and not individual creatures separated from her or in the midst of her, in a superior position ruling the world. My hypothesis is rather that we are in a state of immersion. With a privilege—which perhaps isn't one after all—rather a burden, which is our consciousness, since we seem to know that we know what neither the plants nor the animals know. (The plants and

---

[3] Gilles Clément and Louisa Jones, *Une Ecologie humaniste* (Paris: Aubanel, 2006).

the animals know, but they don't know that they know.) Due to this consciousness, we are responsible—that's where our task becomes burdensome—because we are obliged to anticipate, to construct future scenarios, and to tell ourselves that if we do this or that, there will be such and such consequences. We are conscious of having a particular role in this living world, but that's all. Finally, that gives us a responsibility as gardeners with a friendly concern for the rest of the beings that surround us. But it is true that there are also gardeners who, on the contrary, act as if nature ought constantly to bend to their wishes, as if it were necessary to struggle constantly against nature to ensure their position.

"CAMBIO VERDE": THE BARTER OF WASTE

The simple gesture that we make in our own garden, which consists of making nettle manure to improve the health of the plants we cultivate, the very same that is mentioned in the periodicals *Les Quatre Saisons du jardinage* and *Rustica*, is not so very different from the management methods of urban society when it comes to undertaking an ecological form of management for the reciprocal benefit of humans and nature. That is what Jaime Lerner, mayor of Curitiba in Brazil, a town of 3.3 million inhabitants, put in place more than thirty years ago when he proposed his "cambio verde." The inhabitants of the favelas organize among themselves the selection of recyclable products, and every fortnight, in the square of a favela, in the poor quarters, an empty truck arrives on which is written "Lixo que nao e Lixo," meaning "waste that is not waste." There the people gather with their wheelbarrows; they have sorted the recyclable refuse and they have it weighed. They are given a piece of paper with the number of kilos. Then they go to the fruit and vegetable truck where they are given an equivalent weight in fruit and vegetables. On another day the town will supply pens or books. Jaime Lerner is a town mayor, not a gardener. But, nevertheless, I rank him in the category of planetary gardeners. In Curitiba's municipal nursery the little plants are raised in pots of mate tree fiber, recycled directly without going through any transformation process. The official public buildings are built from reclaimed telephone poles rejected by the state of Parana. Lerner is a politician, he manages a town, and not a small one at that. There is no reason to differentiate between minor and major gestures, to separate the most insignificant of gardeners from one who is the greatest city administrator.

THE ONLY TRULY POLITICAL PROJECT IS ECOLOGICAL

The responsibility of the politician becomes greater the more constituents he has. That is why I believe that today the only truly political project is of an ecological nature. It is hard to see how any society could be organized without this concern, all the more appealing in being innovative and opening up completely new directions. It follows that there must be a high level of awareness at the top, which means, in the government of a country. I spoke with Dominique Voynet when, as minister, she was awarding the landscape prize (a strange prize because it is not landscape architects who make the landscape, but farmers, truck drivers, etc.; we do repair work). Anyway, I happened to receive this national prize. Dominique Voynet deplored the fact that her prime enemy should be her government, because in practical terms, as soon as she tried to do something, all the other ministers were against it. Successive ministries, of ecology, the environment, and others, have always had a spanner thrown in the wheels, minute budgets. It's a kind of façade that provides a clear conscience and comes to nothing, obviously, otherwise we would have long had combined road and rail transport, solar energy, and so on. To give you a personal and frankly trivial example, I can tell you about my own house, to illustrate how behind we are in France when it comes to renewable energy.

For fifteen years I have been trying to install solar panels here (I don't have electricity in my house). I had to turn to a friend, who has already carried out installations, to advise me on the type of inverter, the various materials to use, but also to come in person and help me. Technical help for such tiny operations is nonexistent. We are constantly being dissuaded from undertaking such installations. People speak about the cost being still too high, etc. I have spent 4,000 euros doing this, that's the cost of an ordinary small motorbike that actually consumes energy: for that sum, I can fabricate it. It's not inconsiderable, but in relation to the budget for equipping a house, it's not that much.

However, there is assistance available from Ademe (a public agency concerned with, among other things, energy efficiency and renewable energy projects), provided you are connected to the EDF electricity grid. I contacted Phebus (a local center for renewable energy offering educational courses and advice) and Ademe. Each time I was told that there was no problem, that various forms of assistance and support were available, but they go on trying to persuade you to connect to the network grid. And what if you don't want the network? You have a right not to have the grid, but in that case access to support is impossible. In the year 2000 when we had the storm, I was the only one to have heating. All the boilers in the region run on wood or oil, but they are fired by electricity. It's insane. Why do we have this system? Why does all the equipment rely on a power source that in itself is dependent on a central supply located goodness knows how far away? It's very dangerous. It is understandable that a state should not wish its subjects to become autonomous, but at the same time each one has the right to autonomy if he or she so wishes. In France that's not possible; that is, you have to fight for it.

ECOLOGY STUDIES THE LINK BETWEEN LIVING CREATURES

Ecology, since leaving the field of science, has joined the political field. Politicized, it presents itself to us in that guise. Now, in the mind of Ernst Haeckel, who invented the term at the end of the nineteenth century, ecology was the study of what happens at the level of exchanges between living creatures and their essential relationships. Something dynamic, not a collection of creatures and objects, but the energy that connects them one with another. An overwhelming vision that certain people had predicted earlier (George Sand had already spoken about it, Lamarck before her). Reading Lamarck, the first theoretician of evolution, well before Darwin, one is amazed to see to what extent he was ecologically aware, even though the word didn't exist at that time (1802). In an all-embracing philosophy, he had studied all living creatures, including shells, anthropoids, invertebrates . . . He was able to draw conclusions about the relationships between organisms. Today we know that such and such a caterpillar lives on such and such a plant, which itself may harbor another parasitic insect, that in turn has bacteria or fungi, and so on and so on. Everything is so intimately connected that the "butterfly effect," an expression that has become famous, can be understood in relation to any kind of serious observation (when someone sneezes here at Mouans-Sartoux, that movement of displaced air is not completely disconnected from a typhoon at the other end of the planet). Meteorological forecasts show us the huge anticyclones and movements on the scale of the planet and tell us what the weather is locally. We have this planetary awareness, no matter what happens. This is becoming a familiar story.

Ecology is this thoroughly dynamic dimension, and not something capable of being frozen. There is a nostalgic line of ecological discussion that suggests "in the past life was better"—that is not at all certain—or else, "such and such a landscape is the ideal for living a happy life." What do you do if the landscape changes? In reality, landscape does nothing but change, since it is largely composed of living creatures that are there only temporarily, who transform the landscape. *All life is dynamic; it is constantly inventive.* It is impossible to imagine a motionless and frozen landscape. I willingly set myself apart from those ecologists who see the world of nature as a sort of museum in which humans disturb the calm. For me, humans are an integral part of the system.

WHY "HUMANIST"?

I don't know how the dictionaries define humanism. For me, it is that
which in some way is going to improve human life, allow us to live in
harmony with our environment: a give and take. Humans are taken
into account, integrated in the whole. Nature is not necessarily at the
service of man; he exists within her, submerged in her, and therefore
intimately associated with her. I have to make this clear because it
is not something everyone accepts. The Planetary Garden came out
of this same idea about humanist ecology: it's an integration. That
is what I have been able to conclude, initially by traveling. I realized
that there was a vast "planetary intermingling" that has increased
enormously in the past decades, because human activities mean that
we are constantly moving and making everything else move. Consequently, plants and animals meet in new and unforeseen circumstances that geography does not permit spontaneously. It is one of the
characteristics of the garden, which, historically, is a planetary index.
I was there when the pygmies of Cameroon first adopted a sedentary
lifestyle; it was very moving. Historically, this was how the garden
was born for humanity as a whole. One may suppose that the garden,
being a place of protection for everything considered precious, owes
its existence to humans' progressively sedentary lifestyle. What do the
pygmies do? They go off in search of plants to put in the garden . . . At
a transcontinental level, this is exactly what happened in the case of
the great historic gardens. A garden is always a planetary index. Ecology destroys the notion of the "enclosed" garden. The word garden
comes from the German *garten*, which means *enclosure*, an enclosed
place. With the advent of ecology, people realized that this enclosure,
though essentially under our control, is an illusion. Butterflies, wind,
birds, seeds, even people: everything communicates. If you pour a cup
of bleach down the washbasin, it goes off into the ocean. People knew
this, but they did not know that the life of the ocean depended on it.
It is very shocking, it is one of those things that confront us with a
different dimension and a different scale: the horizon is no longer the

limit of our landscape since we know what is happening on the other side from now on. Our new garden, from one connection to another (biologically speaking), assumes the scale of the planet. Meanwhile, life itself is once again confined within an enclosure. Instead of being limited to a small space that we control, from now on the garden is placed within the limits of the biosphere. There we have a new enclosure. Once we become aware of this reality, it makes us responsible in our role as gardeners, since we realize that life is in our hands. The planet is almost completely gardened. The spaces that are not gardened are known, surveyed, photographed, and analyzed by satellites, and consequently the total surface of the planet can be compared to a place under surveillance, a garden.

THE POLITICIAN MUST BE A PLANETARY GARDENER

In conditions such as these, the politician in his activities becomes a planetary gardener, but he is not necessarily aware of it. This was the subject of the exhibition "The Planetary Garden." If the company administrator were aware of his obligatory mission to be a gardener, aware of being in charge in a "friendly, cautious" way of this diversity that surrounds us, he would act differently. I have been giving examples based on the theme: how can one exploit this diversity without destroying it? Because we are obliged to exploit this diversity since we are dependent on it: we eat nature's produce, we live on nature, everything comes from nature. We can do nothing without the things that nature produces. How should we continue to live in our industrial world, so as to exploit all that without destroying it? Here, the large companies of the chemical industry (such as Monsanto[4] and

---

[4] A multinational company specializing in plant biotechnology that employs nearly 13,000 people, with a turnover of $5.4 billion. This is the company that introduced the chemical Agent Orange during the Vietnam War, the herbicide Roundup (supposedly without harmful effects on nature, an argument for which it was condemned in 2007 for false advertising), as well as seeds genetically modified to resist this product.

others) play a significant role; they exercise a choice, withholding access to this diversity or destroying it: they are murderers.

BIODIVERSITY IN CRISIS

Certainly, there are a lot of people who think they are gardening while actually impoverishing the flora of their environment to keep only a few beautiful rose bushes, a few hydrangeas, the image of a garden appropriate for the individual. But there are also all those in industry who go and take a piece of the Amazon and leave nothing behind but a land transformed into laterite, which is to say, sterilized. . . That is no longer gardening, it is devastation. At any rate, it is not gardening as we are obliged to imagine it today. For a long time, because people believed nature to be inexhaustible, they exploited the earth without restraint (and, unfortunately, they continue to do so). Beautiful gardens were designed solely to be viewed, as paintings and great vistas. They went as far as the horizon which they believed they controlled. There was some justification for this way of doing things. But today, we have to imagine things differently. There is no longer any question of eliminating this diversity for the sake of two or three species that are going to look pretty in the picture: a few rose bushes on a lawn. Close-cut lawn, the most polluting of all groundcover, the most damaging on the planet. We have made some progress in that respect: near Nantes, there is an ecological golf course. But when, for example, I proposed a golf course without grass on a small island in the lagoon of Mauritius, I received a shocked refusal. In a country where it is difficult to get water, where the water runoff is going to contribute to killing the lagoon, they decide to make grass grow, using lots of water that comes from who knows where, at great expense, with lots of chemical products, fertilizers, grass which will have to be mown with polluting machines.

THE DREAM GARDEN

There are ways of growing crops by direct sowing, extensively, over millions of hectares, as they do in Brazil, Australia, even Europe, without traumatizing the earth's surface. There is no need to practice intensive devastating agriculture. A garden is perhaps a little different, because it is not a question of feeding a population, but of a territory where everything is intermingled: flowers, fruit, vegetables. I define the garden as the only territory where man and nature meet, in which dreaming is allowed. It is in this space that humans can be in a utopia that is the happiness of which we dream. That is why it does not need to follow the aesthetic canons of a particular style. In the magazines, they always show the same thing. But it is not necessary to copy that. Better to aim for something that is personally satisfying, that is right for us, and not something that is right according to what it would be appropriate to do in polite society. And in a garden, the concern for an aesthetic resolution is expressed through a way of defining space, so that the result of attempting to interpret the best way of living with nature looks as attractive as possible. You may want a thorny plant, off-putting to others, perhaps, but right for you, the chosen user of the garden.

When you enter a strange house, you may feel good. And then discover a monstrous ornament on the mantlepiece, notice that the tablecloth is rather ugly. In the end, everything is hideous, but you feel good. That is something to think about. What is resolved by asking this question? It may be that you are in a space that is "right." The person who is there has done something that makes you feel good, whatever your taste, and you can tell that that person is happy in this place, and that that can be felt and seen. You feel good because he or she feels good. In which case, you have nothing more to say on the question of taste, good or bad.

THE THIRD LANDSCAPE

A garden is inevitably the result of man's intervention in nature, without it there would be no garden. If there is no gardener, there is no garden. The wild spaces are a space apart that I call the "the third landscape" (*tiers paysage*). In a landscape analysis, commissioned by the center of art and landscape in Vassivière, Limousin, I introduced the notion of a binary landscape: shadow for forests, light for pastures. In both cases, it is a question of managed spaces; it is the engineer's territory: the forestry engineer for the forest, the agricultural engineer for stockbreeding. When I looked more closely for

The third landscape: roadsides.

diversity in these places, I could not find it. It lies somewhere else, in the places that are neglected: the sides of the road, abandoned plots, *friches*, heath and peat bogs, anywhere where it is difficult to exploit the land with machines. It is the sum total of these that I call the "third landscape," a precious assembly, if you consider what it represents as a legacy of genetic diversity.

The wild garden, what I call the "garden in movement," is more an intervention by humans in a territory where they interpret nature, diversity, and the synergy between these phenomena, in order to make a place for themselves, but still without destroying that diversity. As I was interested in the preservation of insects—the most sensitive link in the ecological chain providing very good bioindicators of space—I got to know all the ups and downs of these insects: how they eat, how they are eaten. I arrived inevitably at the ecosystem. As I see it, it is no longer possible to garden without being aware of the ecosystem. I continue to think that way, even if my manner of doing things differs depending on the type of brief.

I was consulted about a three-hectare wood, a project for the park of the château of Mouans-Sartoux. When I saw this wood, I tried to find a way of responding that would improve the whole without really touching the wood, because I found it very beautiful just as it was. I focused on the improvements that might be made to the part assigned to concerts and everything concerned with the buildings, by redistributing certain elements in the space, enlarging certain rides, in short, by redefining this park. For the wood of green oaks, I proposed a double edge, to encourage visitors to walk through the wood, considered to be too gloomy. Elsewhere, I proposed introducing rocks of modest size, similar in volume to a lump already lying under a pine tree, placed there by chance. Each of these new objects, transformed into fountains, benches, or sources of light, placed in a clearing, is inscribed with a word from the French language. These words refer to the vernacular names of plants, grouped together elsewhere in a garden of light near the buildings, a sort of treasure hunt for children and adults.

THAT IS WHERE OUR GENETIC FUTURE LIES

Happily, there are experiments that demonstrate that human intervention is not completely irreversible, as long as species have not disappeared. When a piece of land in a very traumatized environment is left to cleanse itself, it begins by becoming impoverished, because it was loaded with fertilizer. Then, progressively, it will recover a series of flora corresponding to its natural capacity to accommodate diversity: it takes seven to ten years for

the equilibrium to be restored in agricultural soil. Species that
were no longer visible then begin to reappear, having lain as relics
somewhere in those little pockets of resistance that constitute the
third landscape. This is our genetic future. The sides of roads are
extremely precious places. More attention should be paid to them,
because there you find a diversity that has been driven out elsewhere. It is in such tiny examples as these that biological and genetic
resistance can be found. Of course there are species that cannot live
there because the space necessary for their development is insuffi-
cient: a large mammal, such as a deer, will not live there; whereas
an insect, or a mole, will find possibilities. Some species disappear
for those reasons—an imbalance between surface area and required
habitat—others because they have been completely eliminated by
pesticides or brutal fertilizers. Every day species disappear from
the planet. During that time others are discovered: that happens
because diversity has not yet been fully catalogued.

I don't know whether human beings are biologically pro-
grammed; I think not. I think that *life is inventive,* truly, and that
that is the real meaning of evolution. There is a theory that supports
programming. For instance, some Japanese scientists suggest that
an animal or plant species (why not human) could in fact be pro-
grammed and their time determined. This has not been proven, but
it is a possibility. That would mean that all sorts of fossils died out
naturally at a given moment when their program came to an end. I
don't think that is the case with complex creatures that have the
capacity to reinvent themselves; and humans are certainly among
the most complex. It seems that we only use one eighth of our brain.
If that is indeed the case, we ought to evolve in the direction of a
better exploitation of this high-performance organ. If this happens,
our capacity to modify and constantly change direction should
increase over time. Besides, I am not a supporter of the determinism
that says that a person carrying genetic information from birth
would never be able to modulate that profile. I believe that the

impact on human beings of the environment in general, and more particularly of cultural pressure, transforms them during the course of their life. The theory of evolution corresponding to the transformism of Lamarck is not dead and buried. It could reappear in a modern light and provide useful food for future discussion.

The world economy is out of control, and I think that we are going to move toward "deglobalization"—it may even happen without our choice—which means localized consumption and manufacture of goods.[5] That will not prevent our having an economy on a planetary scale, but it will have to be organized at a local scale. That is what we should progressively come back to. This is probably going to happen for strictly economic reasons, and not for ecological reasons alone, since it is the economy that determines everything. The transportation of merchandise and people is already a serious ecological and economic problem on the planet.

Deglobalization will have to be accompanied by a structured reduction in growth. For a long time, those in favor of an alternative world view based on new economies have been advancing the idea of a reduction in the manufacture of goods whose use and fabrication result in significant economic, ecological, and social upheaval on the planet.

The principle of reduced growth implies a decrease in the accumulation of possessions. Material possessions being considered as the riches to which everyone is supposed to aspire, reduced growth appears to be a frustrating principle that does not rouse any immediate enthusiasm. Unless it were possible to find compensation for the decrease in possessions, something immaterial (or partially immaterial) that would be highly valued by a society suddenly enlightened by ecological necessity. The redefining of the biotic

[5] Alberto Magnaghi, professor of territorial planning at the University of Florence and Architectural School, has for many years been coordinating a national research project on local self-sustainable development.

substrata: water, air, and soil form part of those new objectives that will enable these material elements, once redefined, to generate an index of immaterial values compatible with new ways of living.

The question of the model transmitted by a leisured society (rich society) remains a crucial question because it is the driving force behind the desire of the majority to achieve it, no matter what the cost. It must therefore be the driving force of a specific economy: providing access to it.

What kind of model do we want to develop, to allow both the exploitation of diversity for the benefit of an expanding world population and the preservation, in terms of quality and quantity, of that diversity upon which, whatever happens, we depend? This is the question underlying the Planetary Garden, the question to which we must attempt to respond as quickly as possible, if we are not to succumb to the inevitability of irreversible forces of destruction, but rather to develop in every conceivable field a mental territory of optimism—a garden.

The Wisdom of the Gardener

The house at La Vallée.

A GARDEN TOUR

The world of gardens includes gardeners. Without them nothing would exist. But it also attracts broadcasters, propagandists, entrepreneurs, contractors, journalists, and a crowd of scholarly people polished in the art of speaking about it who are called (in French) "amateurs," from *amare*, to love.

The garden "amateur" is not someone who dabbles. He is thorough, travels, compares, makes enquiries, attends exhibitions, discussions and symposia, forms an opinion, constantly refining his knowledge. He is a scholar. Nowadays, the word "passion" is accepted without any distinction as one of an extended range of intellectual pleasures. Amateur, degraded through usage, indicates a nonprofessional category, hence superficial, incapable of getting to the heart of the matter. Garden amateurs are the exception.

The amateur is not necessarily a gardener. A gardener would not know how to be an amateur in terms of his own art. He operates from within. There is no such thing as a gardener-amateur, whereas there are garden amateurs.

At a certain moment, the two meet for a garden tour. The gardener offers his experience, which the amateur immediately files in his personal records. The amateur retains sufficient knowledge (and related dreams) to take on the visit of an unknown garden himself. And to comment on it.

Asked what he sees, the amateur talks about the space and the plant species that are found there. Whenever the occasion arises, he demonstrates his talents as a botanist—there is nothing in the way of garden flora that he is not familiar with. He describes the plants in minute detail, points out their comparative rarity, the difficulty of tracking them down throughout the world, transporting them, arranging them in the right order.

An attentive audience offers the ideal conditions to get him going. If nothing interrupts his speech, nothing pricks the enchanted bubble within which the amateur-guide protects and feeds his emotion, he may possibly reach the topic of History and its development. Then the amateur, extraordinarily enlightened by his inner passion, and darting from one clue to another, discovers within the garden the ruins of Babylon, a sacred hill evoking the memory of Hellenic philosophers, a Mogul pool neglected by Alexander in the depths of a valley, a Mudé-jar portico where Boabdil sighed on leaving the Alhambra, and so on, until his quotations have been exhausted.

Worn out, the amateur will reach his conclusion, happy to announce the date of a future visit: very soon, the garden of Mr. and Mrs. So-and-So. Very exclusive, inaccessible even, a privilege. As an exception, the owners have agreed to welcome us, photographs are allowed provided they are not published, and so on. There will be oohs and ahs of delight. All my sympathy goes to Mr. and Mrs. So-and-So, who, I know from another source, are not entirely put out by showing their masterpiece, insisting nevertheless that "this is the worst moment of the year: you should have come a month earlier!"

In this situation, the amateur is not disarmed. He begins to talk about the weather.

The drought, gusts of wind, tornadoes, frost, clouds, the calendar of saints, constitute an inexhaustible source of devastation and strategy in the garden.

Even though the amateur is an expert on everything, the pattern of intemperate weather is beyond him. The sky and its changes involve the whole of humanity. Everyone can describe an experience in which he has been a victim. The sharing of misfortunes unites human beings in a solid front against nature. Each one prides himself on a fatality, even while admitting that that of his neighbor is not insignificant.

This meteorological exchange brings to an end what is commonly called a garden tour. The crowd, proof against fire or water (according to the weather), breaks up after an outpouring of effusive thanks.

At that point, it seems there is nothing more to be said.

BY FORCE OF CIRCUMSTANCE—through my profession—I find myself one of the amateurs and one of the So-and-Sos. When I have to show people round my own garden, I hold in readiness for the imprudent visitor-amateur a well-rehearsed complaint. The cyclonic storm of the winter of 1999 helps me beyond belief (at the time of writing, the drought of the summer of 2003 promises further assistance).

The western fouchtra,[1] where interesting shade-loving species used to find shelter, today resembles a tangle of thorns. Here gardening is a form of heroism. You have only to mention casually a few survivors of the disaster, as you pick your way Indian style through a thorny stretch, to awaken in your visitors the sense of a shared adventure.

Wearing a bush hat, armed with a pair of pruning shears, grabbing a machete as we approach a climbing rose suddenly become a monster, I propose an eco-garden-tour, inspired by the eco-guide that I become at that moment, casually offering a few useful warnings:

"Be careful of the ford, there are leeches in the water."

"Don't go too near the hogweed, the sap of the plant produces burns at the slightest contact" (the guide willingly explains the photosensitizing effect of the giant hogweed by offering a few examples). Or again, more hesitantly: "The foxglove, like the ivy and the black morel, is extremely poisonous."

The aim being to gently terrify the visitors without their running any risk. Through this dose of information, nature takes shape in all its contradictions, welcoming and cruel, dark and bright, able through one image chosen at any point in the landscape to evoke anxiety and admiration. What interests the visitor is not so much life as whatever threatens it.

What happens after the visit? Does the guide have enough energy left to become a gardener once more?

One sound drives away another: the song of the birds replaces human chatter. Gradually, the animals find their place again and reveal themselves. They had been forgotten. For good reason: they had hidden.

---

[1] *Fouchtra* indicates a savanna-like habitat and is used to describe regions in the Auvergne. The word seems to be derived from a state in western Zambia.—Translator's note.

AND YET, IT SEEMED as if the whole topic had been explored at the same time as the garden. The flora, the style, the architecture, the ornament, the light, the water, the passing of time, the current weather, nothing seemed to have been left out of the description. Garden as object, intended as a painting for the onlooker. An amenity area for the person who looks after it as a neatly tended plot. The legitimate extension of an immaculate house. The other inhabitants of this environment under surveillance, whether visible or invisible, never seem to have any rights.

Garden books don't mention wild creatures, except to explain how to fight against them.[2]

There is never any question of their being natural inhabitants. Publications remain obstinately silent about the moles of Babylon, the beetles of Villandry, the dragonflies of Versailles, the snakes of the Alhambra. And yet, they must still be living there. However, none of them have any bearing on the artifice for which the gardens are famous. Tradition excluded from the gardened area all those living species, both animal and vegetable, that escaped the control of the gardener. The vagabond has no place there.

The advent of ecology destroyed this vision. Its primary concern is nature in its entirety and not the garden in particular. However, the garden is made from nature. Birds, ants, mushrooms, insects, and light seeds recognize no boundaries between territory that is policed and space that is wild. For them, anywhere can be inhabited.

The constant influx of mobile species represents a considerable force against which the struggle to garden is transformed into a war. There is no lack of arms. They form the mainstay of shops professing to defend the garden, but which in reality attack it. In prime position on the shelves, a terrifying range of anti-mole products, followed by variously colored powders aimed at the eradication of ants, field mice, slugs, aphids, red spiders, mealybugs, ladybirds, eelworms, and so on.

---

[2] With the exception of specialized ecological journals, such as *Les Quatre Saisons du jardinage*.

In the garden of my childhood, you had to obey the rules: follow implicitly the commercial instructions. We had to smoke out, spray, burn, weed, find every possible way of treating rebellious nature, so disastrously inventive.

I had learned to intimidate the moles by sticking bottles into the ground, their bottoms broken off, the neck exposed to the wind in such a way as to produce sounds that drove the animal away. The bristling lawn, turned into a minefield, was the cause of a few accidents. It particularly attracted the sarcasm of supporters of the Greens who saw in this masterpiece a pathetic version of *arte povera*.

In addition, a few shards of glass were placed at strategic points, which the mole, being skilled in bypassing, would avoid. Hemophiliac, reputedly dying from the smallest cut, it could not possibly escape alive. Yet we never found the corpse of a bloodless mole. As for the fragments of glass, they would rise to the surface, sprinkling the grass with glittering colors.

The hosepipe technique, costly in water, reveals its limitations to anyone who hopes to use this means to drown the subterranean animals. Signs of their reemerging at some distance from the flooded hole quite soon reveal the transport capacity of the network. Immense, hopeless.

In my father's garden, smoke rockets were used only once. There was a rumor that the smoke, related to mustard gas, had led to two gardener-arsonists being hospitalized at Guéret. After experimenting with the most varied baits—among them, the earthworm kit, an unappetizing brownish gray, fresh from a toothpaste tube that had to be squeezed while taking care not to touch the product with one's fingers (the "human" smell, people said with knowing looks)—we were convinced that only serious poisons, vouched for by the great poisoners of history, would allow us to achieve our goal: to exterminate the moles.

Killing moles with strychnine requires experience and patience.

We carried out the proceedings with the utmost rigor. In order to kill moles, first kill worms. The captured worms died in a tangled mass, writhing in pain. Anyone who has taken part in this double murder knows how much time and mental energy it requires, causing the gardener deep misgivings. If one were a mole, the last thing one would want would be this red and blue gel composed of inanimate worms. What was going to happen?

All these experiments led to a temporary reduction in the number of molehills on the lawn around the middle of summer, without it ever being possible to determine whether that was due to our efforts or to the drought. During the warm season, the animal digs its tunnels deeper, and goes off toward the woods and the wetlands where it finds its food. Whatever the case, we maintained a daily relationship with the mole that created an intimate bond. Furthermore, when we unfortunately happened to catch one, we were overcome, dumbfounded by such a pregnant victory, and close to tears on discovering that the animal was no longer moving. There is still in us something of the cat playing with the mouse, as long as the mouse is alive.

It is impossible to look at every aspect of the mole question, there are so many products and methods available. Just one last example, to conclude this list: the shotgun. At certain specific times—in the morning, at midday, and in the evening—being ready to fire into the molehill or to one side, it doesn't matter: the explosion produces a cardiac arrest in the animal. Sometimes, in the huntsman too. I have witnessed the following spectacle: a gardener/military man jumping through the kitchen window, his gun in his hand. He had seen the earth move . . .

Every species declared harmful generates hoards of murderous inventions. The gardener, convinced of his right to eradicate, wallows in a paranoia actively encouraged by the poison sellers. He becomes a slave to a complex, useless, and harmful practice. Anything that is not part of his "project" must be removed from the landscape. Animals are a nuisance.

When I was able to acquire a piece of land, the question arose: is it possible in this place, which is sufficiently remote to attract

wildlife, to combine a garden with nature itself? To establish a shared territory? Would the animals get what they wanted out of it? Would they accept my presence? How would one re-tame a fauna expelled for so long?

With no example around to help me, I had to experiment. I had decided that one part of the garden—being biomass: leaves, fruit, rhizomes, seeds, etc.—would be given back to the animals accustomed to eating it. That meant giving up a portion of space.

Without wanting to go that far, but wishing for it all the same, slowly and without any precise plan, I made a garden for the animals.

A house too.

No informed planning. That's just how things happened. The garden tour, if I had to stick closer to reality, would begin with a description of habitats: places made for the inhabitants, the animals. Some of these inhabitants have become familiar. They have been given names to which they are not expected to respond: a simple means of recognition among a crowd that I know to be innumerable.

Leopold, the boldest, makes his appearance early in the morning and two hours before sunset in summer. He is an adult roebuck (*Capreolus capreolus*). He crosses the bank in the west to reach the giant hogweed (*Heracleum mantegazzianum*) in the shade of the Japanese dogwood (*Cornus kousa sinensis*). He strips the rough stem of the plant by rubbing it with his horns to reach the pulp, tastes it, and moves on. Here and there he samples a mouthful of foliage without ever lingering. Madame Leopold and the four fawns, more timid, carry on in the same way. The passage of roe deer—a single family for a garden lost in the forest—leaves few visible traces, except in the winter when the animals are looking for the tannin in the young trees and gnaw the bark. The terrain round the house, of which I keep a record, can probably not tolerate a greater number of these elegant and rustic deer. They sleep on the ground itself, without any shelter or den. I find beds of trampled grass in the field between the planted trees that enclose this space on its fourth side. The three other sides are made up of the tall, ancient forest edge. They transform the meadow into a pasture, a safe place where game come to browse. It's not unusual to find Leopold there at dusk among the abundant fennel whose taste appeals to him. If one stands watching him, he barks, leaping into the air like the impalas of the African savanna. If one stays, he describes a large circle round the observer, from time to time uttering his cries, in a vain attempt at intimidation; he is making the point that the field is his domain.

Gaston belongs to a dynasty of nutria (*Myocaster coipus*, also known as the water rat) nesting among the roots of an old oak in an

House under construction, La Vallée.

overhanging part of the bank. Placid by nature, he travels through the water, taking his time. At the sight of a pedestrian, he slows down even more, cautiously weighing his possibilities of escape. But Gaston is curious. He comes so close that one can count his whiskers. He passes as if nothing was going on, indifferent to anything that does not concern his chosen milieu: the water, where he moves effortlessly, creating no waves; barely a ripple on the surface, a fleeting trace bearing no relation to the imposing mass of his body. In reality, he is observing the observer and distrusts him. The direction he takes is nothing more than a lure to deceive the enemy as to the exact address of his lodging. With great patience, hidden by the bushes, and keeping absolutely quiet, one can see Gaston turn back, climb up the bank, shake himself, and return home. At dusk, I have seen him showing an interest in the festoons of greenery above the creek. The stepped bank, sculpted like a cliff by the regular lapping of the waves, presents an obstacle to reaching the top. He has had several attempts at climbing this oversize step, but his heavy body always pulled him down. He finally found a passage between the heather and the velvet grass (*Holcus lanuginosus*). Then he began to climb the hill, a strange destination for an aquatic animal. Stranger still, the cry he uttered at regular intervals, indistinguishable from that of a duck. The classic texts say nothing about the nutria's complaint. Was he complaining, was he calling for a lady Gaston? The following day we found his corpse two meters from his principal habitat. Gaston was entitled to a simple dignified burial, an informal gathering on the shaded hillside that borders the river, his territory.

The life of a nutria is not very long. Their corpses, appreciated by vultures, carrion feeders, and other burrowing insects, quickly disappear. Sometimes, one rediscovers only the head, with its two large, well-worn, burnished teeth. I found one of these remnants under the ridge of the roof where Décibelle hides her pickings. Décibelle is a noisy, smelly stone marten. She burrows inconsiderately into the roof space between the laths and a layer of insulation, ideal for her to slide down . . . Since the arrival of Décibelle, the small rodents pottering about over our heads have disappeared. Shrews, mice, bank voles, all are food for the Mustelidae family, to which the marten belongs. Nocturnal in her habits, she starts moving at dusk and returns in the morning around five o'clock. Several times during the night, she can be heard clearing a passage through one end of the roof into the tunnels of glass wool. She comes to feed her babies, whose urine filters through the ceiling. Time and again, I have had to get angry with her, using the handle of a broom to strike the ceiling beneath the nest. Décibelle has moved house toward the bathroom where she no longer disturbs those asleep.

Her visits have become fewer: there is competition.

In the roof, Edouarda lays down the law.

On 18 May 2003 at 11:00 P.M., a violent thumping that the noisy Décibelle could never have been responsible for reverberated directly below the chimney flue. The roof cavity is not sealed around the flue. There are gaps. Armed with a torch, I try to see what agitated animal is the cause of the thumping. Through the largest of the openings, the coil of a snake with a light belly is slowly passing. The impressive

Roof of the house at La Vallée.

diameter of the body corresponds to an animal approaching two meters, a normal length for the adult *sanghyar*, otherwise known as the green and yellow grass snake.[3]

According to local legend, the *sanghyar* did not originate in the Limousin. It is believed to be an exotic serpent, parachuted in over

---

[3] *Sanghyar,* vernacular name for the green and yellow grass snake seen in the region of Berry and the Limousin Marches (*Zamenis gemonensis Laurenti,* or *Zamenis gemonensis Viridiflavus*).

the region in the 1960s to get rid of the vipers that infested the countryside. Witnesses agree on remembering airplanes passing above the forests and heath lands. What didn't go on in the 'sixties!

The green and yellow grass snake is part of the indigenous fauna. Vindictive, aggressive, it is one of the largest European ophidians along with the Montpellier snake (*Malpolon monspessulanus*). Rollinat describes it as resistant to any form of domestication, capable of biting and refusing to let go.[4] He carried the marks of such bites, harmless but painful. The captive creature, kept separate from the other reptiles studied in his laboratory at Argenton-sur-Creuse, was as savage at the end of its stay as at the beginning. In the mating season, the couple's antics come to resemble a boxing match. So I was the astonished witness of a rooftop copulation. A lot of things go on in the roof. That evening I didn't attempt to discover the partner. With the help of my all-purpose broom, I stood on the piano and invited her with a sharp blow to go back to the slopes and move away from the hole. Immediately, Edouarda reacted by showing her head. It was indeed the sanghyar, bold and powerful, whose shed skins lie about all over the garden, in the lea of the house, but also in the climbing hydrangea and the black-fruited vine.

So there was an explanation to the endless chasing and sliding heard the previous night, the leaps and bounds, cries and blows, followed by the fall of a body into the little pool, which turned out to be Décibelle, every hair soaked, fleeing by the light of the moon, pursued by the ardent serpents.

Rags soaked in turpentine calmed the creatures' nuptial ardors. The next day, I came across an enraged Edouarda. She was making her way up through the woods.

The double roof skin in houses creates a temporary habitat much sought after by grass snakes. Insulation produces practical lodgings for a host of animals, something that Rollinat, who died in 1931, was

---

[4] Raymond Rollinat, *La Vie des reptiles de la France centrale* (Delagrave, 1934, 1937, 1946; rpt. Société Herpétologique de France, 1980).

not able to describe. The Aesculapian snake does not stay there long, preferring the dry woods and rocks, but she comes back regularly.

When building the house, I didn't imagine to what extent it would become a nesting box.

For the last ten years, the double lintel over the entrance door has provided shelter for a litter of Pipistrelle bats. In the spring, their tiny black feces litter the threshold. From the month of June, the young bats fly to and fro from this section of the building toward the garden. Swift, barely visible, too numerous, they have not been given names. Any more than the blue tits, pipits, tree creepers, goldfinches, or thrushes, whose nests occupy the slightest hollow in the walls of the house. Or even the green lizards, squirrels, or traveling hedgehogs. There is space everywhere: between the badly pointed stones, on ledges under the overhang of the roof, in the cavity where furring strips raise the roof slope to divert the flow of heavy rains.

I am describing the most visible animals. But there exists a substantial world whose presence eludes us, composed of creatures that are silent, imperceptible, slow, sometimes mimetic, not seen by our predatory eyes, but aware of one another, avoiding and pursuing each other. This world exists alongside us, without our being aware of it.

Insects, the excitement of the fields, the sharp or humming background, tiny specks of dust driven away with a flick of the hand. An annoyance. Uncut pearls, diadems.

The place where the garden is being developed is a valley sheltered from the winds, where I used to come as a child looking for beetles, butterflies. A few cows used to graze at the bottom after the harvest. The constantly damp soil near the stream is never affected by drought. Higher up, where the house now stands, there was heath land grown wild. Already a few oaks. I bought this rough uncultivated piece of *friche* after it had been abandoned for fourteen years. The dry part, transformed into an oak and hornbeam plantation of middling growth, still offers protection for the butterflies and rose chafers (*Cetonia aurata*) that I had always known. The pesticides spread in the surrounding fields have not destroyed the prevailing diversity. For me, success lies, not in the work of the architect in the arrangement of forms, the balance between shade and light—I am no judge of that—but in the acknowledgment of a life saved.

Better still: the Field. A recent acquisition, only eight years ago at the time of writing. An almost flat surface, approaching a hectare, exposed to the full light above the valley. In this place, transformed into a browsing ground, there grows a mass of multiple herbaceous species. Some of them are the result of seeding at the beginning. The meadow of that time, composed of a single grass (*Dactylus glomeratus*) destined for winter fodder, transformed into a flowering field of varying moods according to the years and the seasons. A regular survey records its state of evolution. The evolving list of plants is worth noting,[5] but the most surprising thing is the enrichment of animal species. This summer 2003 in mid-July, I counted on the buddleia fourteen different species of *Lepidoptera*, which included the scarce swallowtail, a supreme glider, the white admiral, the painted lady, the velvety cardinal, a swift clouded yellow, a versatile swallowtail, whose caterpillars feed among the numerous fennels, a cloud of graylings, and two diurnal hawk-moths: the hummingbird hawk-moth and the elegant bee hawk-moth.[6]

For the tour of the garden—if I did it according to my priorities: animals first—I would finish with the Field, explaining why, of all the sites visited this is the most important.

What are we looking at if not a *friche*, its natural impetus held in check, recycled each year by an autumn scything? A basic gardening

---

[5] An exhaustive list of species observed between 1994 and 1998 can be seen in *Le Jardin en mouvement*, 4th ed. (Paris: Sens & Tonka, 2001). Between 1998 and 2003 the distribution of species has greatly changed. The soapwort (*Saponarea officinalis*), spiked speedwell (*Veronica spicata*), scabious (*Scabiosa columbaria*), greater knapweed (*Centaurea scabiosa*), and fennel (*Foeniculum vulgare*), have assumed much greater importance.

[6] The species mentioned by their common names are, respectively, *Iphiclides podalinus*, *Limenitis anonyma*, *Vanessa cardui*, *Pandoriana maja* (syn: *Argynnis Pandora*), *Colias croceus*, *Papilio machaon*, *Hipparchia semele*, *Macroglossum stellatarum*, and *Hemaris tityus*.

technique, a late drama: nothing is cut before the second week of September.[7]

Our floral diversity, essentially herbaceous in our climate,[8] needs light to reach the soil. There, the plants and their hosts take it in turns: crickets, locusts, grasshoppers—the natural source of a constant stridulation—pearly *diptera*, red leafhoppers, paper wasps, huge ticks, beetles. . . . Between the mulleins, mallows, and evening primroses, visited by the large hawk-moths (*Sphinx ligustri, Herse convolvuli*), is stretched a network of webs where the striped argiope and other spiders (*Argiope bruennichi*) reign supreme. Food is plentiful. The trails of field mice spread out beneath the grass. Kites observe closely from above, and the buzzard, and the brown owl by night. And Leopold, the ultimate proprietor, who tours his terrain in leaps and bounds.

In this vague mist, with no visible paths, see how many paths are permissible. After all, it's only a field.

[7] The mowing takes place between the end of September and the beginning of November. It was Philippe Darge, then president of the Federation of French Entomologists, who recommended this late cut. Observations carried out on the corridors of diversity in forests managed by the Office National des Forêts in Burgundy indicate that this timing of the cut is the best safeguard for entomological fauna that tend to bury themselves at the approach of winter.

[8] In tropical climates diversity is primarily expressed in woody plants. A tropical garden profits from being suspended, indeed "canopied."

MASTERS AND MESSENGERS

My masters are not honorable scholars or simple human beings.
I don't know enough innocents to win my confidence. Except by
looking sideways, beyond the polished realms of polite society.
There I look into autistic eyes, trying to establish a connection,
however tenuous, between my own stuttering and the very
real affliction of timid children, angels abandoned to their own
resources, enclosed within their muteness, indomitable dreamers,
bears.

Such people have no nation. They seem to have no origin, they
travel through time, instinctively questioning, with only their
image to offer by way of communication, as if they were perma-
nently deprived of speech.

Their case is pressing. Nature resembles them.

Pride, presumption? An extract from "Sleeping Notebooks"—a series of notes taken over the course of time, accumulated in exercise books of different sizes and colors that, once completed, fall asleep. I could write this text again, simply changing a few terms. How should I modify it, so as to mitigate the charges of misanthropy leveled in the original version, a position incompatible with "wisdom." For a long time, that was my position.

Over the course of time, I cannot say that I have been exactly invaded by wisdom. I see the wise man—like the artist and the fool, both unattainable positions—floating above everyday contingencies. For the moment, I share with the majority of my fellow citizens the burden of submission to the laws of gravity

Yes, I try to listen to plants and animals, insofar as I can. However, it would be impossible for me to reach them without approaching them by name—which is why, beyond those silent masters, I owe everything to the messengers: teachers, scientists, philosophers, whose role is to astonish us by showing us the fragile diversity that the garden[9] supports.

[9] "Garden" should be understood as the planet (see "The Planetary Garden" in this volume).

There are numerous messengers, sometimes modest, sometimes famous, always close to self-effacement. That distant gaze, that absent-mindedness, enables them to pass on a knowledge that—do they know it from the beginning?—does not belong to them.

All those who like me took the course in plant ecology at the School of Versailles[10] in the 1960s and 1970s will remember Jacques Montaigut. Tall, indefatigable, he succeeded in turning the systematic study of plants, recognized by everyone as the most austere of disciplines, into a book of stories and legends.

It was not enough for the world to be fabulous: it was organized. Difficult to detect the plan of such an organization. Why should the group of herbaceous, mobile gamopetalous[11] plants be more successful or more widely distributed throughout the planet, or more developed than the assortment of dialypetalous plants, hardy but disastrously arborescent, fixed to the ground? Each species forged a way, an ingenious, unique, sometimes complicated way to position itself in space and time, to ensure its perenniality, to survive bad weather, catastrophe, predatory behavior, disease, parasites. Each one invented its way of life. To adapt to the changing conditions of the environment: all sorts of discoveries, useful and absurd, vital and ephemeral. Everything was urgent but everything was changing. With each modification of the environment, how much energy was needed to find a new solution, to invent the future?

Nothing in nature could be seen in any way other than in the light of evolution.

Jacques Montaigut was enthusiastically generous with his knowledge. This contagious pleasure went straight to our hearts. In the rational curriculum of engineering studies, the botanical outings with him took on the appeal of holidays: adventures at the edge of a

---

[10] The École Nationale Supérieure d'Horticulture subsequently relocated to Angers and became ENSH.

[11] Gamopetalous: with united petals; as opposed to dialypetalous: with separate petals.—Translator's note.

cuesta[12] in the Ile-de-France, where the tiny *draba verna*,[13] inadvertently crushed underfoot, prepared to reveal its secrets. A bright outcrop where sheep come to graze in the shade of a blackthorn hedge, overlooking a green valley, banal compared with the dry slope where we had arrived. Growing side by side were common restharrow (*Ononis repens*), military orchids (*Orchis militaris*), field scabious (*Knautia arvensis*), laburnum (*Cytisus laburnum*), Astragalus (*Astragalus monspessulanus*): northern fugitives from a Mediterranean flora. Yes, the southern slopes of these ridges offer a home to heat-loving species. They are found even further north on the spoil heaps of Valenciennes, where the disturbed soil creates a microclimate. The vectors of this intermingling: the wind, the birds, the sheep whose wool contains entire landscapes, all the seeds gather there. Within a few minutes, we were transported to the antipodes, to the southern lands of New Zealand where people say that the imported sheep had introduced European flora and the terrible gorse (*Ulex europeus*), used to make enclosures and become invasive. The evening surprised us, without our having exhausted the square meter of limestone meadow—burnet (*Poterium sanguisorba*), cinquefoil (*Potentilla anserine*)—where we had undertaken to reveal the universe.

To popularize does not mean distorting knowledge in order to make it popular, but rather to express in simple terms the complicated adventure of our planet and its inhabitants. Certain people have this talent. A scattering of written words, the texts capable of making the reader intelligent, are drip-fed onto the market.

I believe that greater danger lies in withholding the truth than in leading people to it. Invited to dinner, I found myself sitting next to a chic and earnest lady. We spoke about traveling, one way of coping

---

[12] A cuesta is a hill ridge having a steep scarp on one side and a gradual slope on the other. The cuesta in the Ile-de-France marks the limit between the tertiary plateaus of the Parisian basin and the limestone plain of Champagne—excellent for growing grapes.—Translator's note.

[13] Spring draba (also known as nailwort, shadflower) with leaves collected toward the base. The plant is never more than one centimeter high, not counting its flowers.

with the distressing shortcomings of such social gatherings: the butterflies of Kenya amazed her. Unable to find words to express her rapture
at such a spectacle, ecstatic, she searched every crevice of her memory.

By the cheese course, the conversation was dragging on the subject
of her inexhaustible garden. She had a horror of caterpillars. I ventured
to say that nonetheless those caterpillars produced pretty butterflies.

Silence. Mask. Change of subject. The embarrassment was mine.
She didn't know. We could have laughed about it, but no, there
comes a moment when innocent conversation becomes rudeness. I
had crossed that line. What light-hearted topic could we broach?
Knowledge demands a seriousness that tarnishes the surface of
things. It is more acceptable to see the world in a bright light and be
grateful, without trying to uncover the origin of its delights. An attitude accompanied by a cocktail party technique: any discussion suspected of being serious is immediately blocked.

In spite of everything, the messengers persist. They never raise
their voice to the point of attracting the largest audiences, but they
materialize from far and wide, hesitantly placing their signature on
the evolution of human thought. Certain ones leave their name in
dictionaries, others are forgotten. All of them have undertaken the
difficult task of observing closely in order to understand.

In my work as a gardener, I keep a few well-known minds high-
lighted above the workplace: Tournefort, Linnaeus, Laborit,
Lamarck, and other more intimate friends jumbled together. I refer
to them whenever necessary, not knowing at that moment whether
I owe them or whether I am giving them something.

Having observed the recurrent wildfires in the South African
Cape region, I adopted a Lamarckian approach to pyrolytic flora. The
conditioning of certain fire-resistant attributes—the fruits of hakeas, eucalyptus, proteas,[14] and numerous seeds requiring thermal

[14] Fynbos (or South African scrub) offer protection to numerous proteas that have
adapted to fire. The hakeas and eucalyptus, naturalized in this landscape, are of Australian
origin, considered invasive plants.

shocks in order to germinate[15]—indicates an adaptive evolution in response to the fires: a slow process comparable to what Lamarck calls "transformism." This theory has been contested by the majority of scientists, whose researches, widely divulged by the English-language media, refer exclusively to Darwin. Recent work carried out by the Swedes, on the transmission of acquired characteristics in relation to obesity in humans,[16] all of a sudden revive the Lamarckian position, or at least make it acceptable as a possible mechanism in the evolution of living organisms. The small Lamarck Garden at Valloires—an extension of the gardens created in 1986—was opened in October 2003 based on a design by Miguel Georgieff, with whom I collaborated.[17]

The man who recognized the importance of clouds to the point of setting up a means for their classification, who invented the term "biology" for the study of living organisms, thereby preceding all ecological thinking, who proposed a theory of evolution fifty years before *On the Origin of Species* (1850), deserves a much bigger garden. A park, a monument. Better still, a piece of nature. At the moment, the chevalier born in Bazentin in the Somme has to content himself with a métro station in Paris, a school in Albert, and the Valloires garden—the initiative of Vincent Delaître, its director. The philosophy of a humanist scholar, transmitted to us by his best biographers,[18] is imagined as a gardened space. The whole of the Jardin des Plantes could be dedicated to this work and it would not be enough. It would be in the interest of public health to make the great men of this world read some of the writings of Lamarck, a messenger who became a master. They would find in them an

[15] Certain seeds require a chemical shock. The smoke brings Restios (large colorful grass relatives) out of dormancy.

[16] The characteristic "obesity" acquired during one generation would reappear as an inherited characteristic in the grandchildren, skipping the intermediary generation.

[17] The Garden of Valloires in the Somme, Vallée de l'Authie, near Abbeville. The Syndicat Mixte d'Aménagement de la Côte Picarde (SMACOPI). Projects directed by Jean-Christian Cornette. Site supervision: Olivier Baert and Miguel Georgieff.

[18] Yves Delange, *Lamarck* (Arles: Actes Sud, 2002).

alternative to the projects repeated over and over again on the depletion of resources. The journal *Hommes et plantes*[19] carries on every opening page a quotation from the great naturalist, taken from his *Système analytique des connaissances de l'homme,* published in 1820:

> Man, through his egoism unable to perceive his own interests, through his inclination to enjoy everything at his disposal, in a word, through his lack of concern for the future and for his fellow beings, seems to be striving toward the annihilation of his means of conservation and the very destruction of his own species.
>
> By everywhere destroying the great mass of vegetation that protected the soil, in return for objects that satisfy his momentary greed, he is rapidly increasing the sterility of the very soil he inhabits, causing the springs to dry up, driving away the animals that found their subsistence there, with the result that large parts of the globe, previously very fertile and highly populated in all respects, are now arid, sterile, uninhabitable and abandoned . . . One might say that man is destined to exterminate himself having made the globe uninhabitable.

This is not a warning. It's a statement. Not so much as a comma to be changed, except perhaps that slight wavering, the result of uncertainty. Yes, the human species is working toward its own destruction, no doubt about that. All the examples, all the leaders, and today all the citizens are aware of the absurdity of the way of life produced by the market economy. At no point does the question of changing it arise. Power is not in the hands of those who are trying to modulate the system in order to delay its collapse.

The human project, whether conscious or subconscious, can be defined in a few words: to die buried beneath its riches.

---

[19] *Hommes et plantes* is published by the Conservatoire des Végétales Spécialisées (specialized plant collections) three times a year.

Fatalistic views, such as the Darwinian position—evolution through natural selection: nature invents, the environment sanctions—affirm that everything is predestined; the strongest survive, the rest perish. Those best built to endure will endure, the weakest will lead a transitory existence. Nothing can better appeal to ambient liberalism: not to encumber ourselves with handicaps, slowness, and delays. We must move quickly, succeed. Succeed at what?

The transformist position, on the other hand, leaves the door open to possibility. During the course of its life, the living organism, whatever it may be—whether plant, animal, human—has a chance to modify itself (either by its own volition or through exterior pressure); it can transform itself. This transformation, once registered, is transmitted to subsequent generations. For the human being, the "conscious animal," this provides the basis for a project, a mental territory of optimism.

A garden.

When Laborit formulated the principle of biological information—the living organism receives a message, interprets it, and transforms it by making it more complex—he gave the "project" the vertiginous dimensions of the unknown.[20] The most efficient information systems are unable to predict the response from the environment, whatever the question may be. Such is the inventive capacity of nature.

When Francis Hallé discusses the delicate comparison between the animal and plant worlds, he reveals the unexpected plasticity of permanently fixed organisms.[21] A priori, vulnerable: trees cannot run away. The stress caused by any brutal modification of the environment (drought, plunder, parasites, etc.) can provoke a modification of the genetic fingerprint in those parts of the plant that are

[20] Henri Laborit, *La Nouvelle Grille* (Paris: Gallimard, Folios Essais, 1974).

[21] Francis Hallé, *Éloge de la plante* (Paris: Le Seuil, Science ouverte, 1999). Hallé is chair of the Tropical Botany Department at Montpellier and founder of the treetop raft project. See also "The Planetary Garden" in this volume.

alerted: recent shoots carry this information and transmit it to the seeds, the future generations (a Lamarckian process, although the term is never used). Plants—principally trees—have the capacity to modify their genome in the course of their lifetime. Animals preserve it throughout their life. Human beings, too. At least, that is what is thought. At this point in time, research offers no other conclusions. The plant world has gone in for genetic manipulation since the dawn of time. There seems to be no reason why the rest of living organisms should not do the same. In themselves, genetically modified organisms (GMOs) and their creation share some of the characteristics of evolution. Our everyday landscape is nothing other than a tracery of genetically modified organisms, combined more or less by chance. If we are to have a serious discussion about genetic modification, we have to consider nature in its entirety. It is not the manipulation that is the problem, but the anthropocentric orientation given to genetic modification and its use.

Francis Hallé takes the time to explain. A teacher's habit.

How can "the influence of the length of daylight on the behavior of humans living in a tropical zone," a tough subject seemingly intended for specialists, be transformed into an accessible text? Everything is there: an appealing title, *A Land Without Winter*, a clear style, plenty of images. The author illustrates his proposal with references drawn from multiple sources: literature, art, poetry.[22] The science seems to be drawn from such interactions, rather than dominating them. It seems to be embedded in a compact whole, capable of being read differently each time. Sometimes one style, sometimes

---

[22] *Un Pays sans hiver* refers to the region between the Tropic of Cancer and the Tropic of Capricorn. In the book, Francis Hallé develops a bold thesis: the length of daylight in the tropics, constant throughout the year, promotes a cyclic weather pattern. Anything is possible at any moment. Outside the tropics, a linear weather pattern unfolds in seasons that are favorable or unfavorable for certain activities (synchronism). An example that summarizes his thinking: in the tropics there are revolts, in other places revolutions. Photoperiodism in plants (the response of a plant to relative lengths of day and night) is well known. In animals and humans, the pineal gland, an upgrowth from the optic thalami of the brain, and long remaining a mystery, is a receptacle for photoperiodic information.

another, depending on the aspect to be described. Occasionally, the measured words of philosophy are appropriate, at other times the metaphors of poetry are better suited, or even the academic language of science. Sometimes, words are not relevant. Drawings suffice.

After eight years of intermittent correspondence, impossible meetings, interrupted conversations, we finally met in the heart of an African forest, in the tree canopy. A successful meeting at a campsite lost in a loop of the river Makande, in the heart of Gabon: the treetop raft project.

The long-awaited moment. Gorilla jungle. A spacious clearing. On that day, assembled at the Makande campsite, I found a new type of person. Unaccustomed to associating with researchers, I discovered people entirely focused on understanding the world and how it functions.[23]

There I met people absorbed by a dream, accustomed to conjuring up the invisible, exploring hypotheses, in order to support a fragile edifice, a theory. That evening, the American, Peter, was giving a lecture for the benefit of fifteen other scholars and an animal population attracted by the lights: insects, bats, night birds scattered in every direction, and nearer the banks of the Makande bordering the camps, more retiring and prudent: crocodiles, snakes, antelopes, forest elephants, monkeys in the trees . . . The slides followed one another on the improvised screen pasted with nocturnal moths, hawk-moths and clouds of flies. Peter's studies demonstrated the incidence of daylight on plant organisms between the top and bottom of the canopy (the foliage of the large trees of tropical forests). He ended his presentation by indicating that these experiments, carried out over the course of several years, did not offer any conclusion. Everybody seemed delighted.

To stumble without hurting the ground, to expose one's uncertainty for general consideration; to test hypotheses, by grinding them in the benevolent cog wheels of the forest. Anything could be said and contradicted. Nothing else mattered because nothing mattered more. I didn't know what to think about the refinements of the situation. Did the overwhelming happiness of working together to disclose the nature of the universe hide some monster? And what would be its name?

Each time a different lecture. So much knowledge in so short a time about such a small area.

---

[23] In the question of diversity, the stakes are enormous. I am not unaware of current laboratories and sponsors, but we were beyond any commercial project, outside their control, so to speak.

Treetop raft, Gabon, 1999.

In its church, the scientific community functions according to a model quite different from the profane. The course of a day, its architecture, its objectives, seem to hinge on what gives man his intangible exclusivity: his spirit. The grandiose project of a group such as this—unfathomable, like that of any society devoted to its dreams—does not need to be formulated to make it acceptable. Nor even to make acceptable the inconceivable energy invested in a place such as the Makande, remote from any form of civilization. No amenities. Better than that: hammocks, litters, stretchers, perched in beds of foliage, and, to arouse the spirit: a container of provisions, an inventive range of food, within everyone's reach. Outside the distribution circuit, although always shared, a consignment of bottles of whisky,

said to be vintage. So much gray matter. So much matter. So much intoxication. I felt as if I had the privilege of sharing the limbo of science: as if I were living in a country far removed from ordinary affairs.

The researchers occupied the terrain. With them, it was possible to climb the trees, open up paths, reach the underneath and the top of the foliage. At every possible height, suspended on ropes, supported by bubbles, perched on the raft,[24] wedged in an "icos"[25] or again on the ground in a lab, they were studying. Early in the morning, between six and eight—the time when the hot-air balloon could navigate safely—two or three human beings could be seen flying over the forest, seated in a triangular gondola: the treetop sled.

I still have an all-enveloping memory of my first night spent on the raft. Indecipherable.

To expose yourself at the top of a tree, with no other horizon but a sea of treetops, no matter where you look, is unlike any human activity. Or perhaps, very long ago. An ancestral memory? A childhood dream?

An animated silence, animal noises. Sounds coming from every direction. And also from underneath.

Clouds. Storm. To be so near the lightning. The rain teems down, abundant. It beats down with all its weight, making that illusory protection, a soaking K-way jacket, stick to our bodies.

The night is spent drying in the moonlight. The foliage is gleaming. The "ra dodo,"[26] an improvised bed, a sort of shaft (for harnessing horses to a cart) on one of the usable inner tubes, has turned into a swimming pool, and is useless. The only thing to do is to sit astride the structure, with your feet on the net above the void.

---

[24] The raft, placed on the trees, assumes the form of the canopy. The initial model, with six inflatable beams and six regular sides, has become an elongated pretzel.

[25] From icosahedron, a polyhedron with twenty sides, a structure with a metal frame, placed in the fork of large trees.

[26] *Ra dodo*: informal term for a sleeping raft, a pun on the French words *radeau* (raft) and *faire dodo*, a childhood expression for falling asleep.—Translator's note.

In the morning, left alone, I wait for the sun and the animals that come with it. Instead of that, I hear a humming, distant at first, like the uneven breathing of a celestial monster. It approaches, invisible. Suddenly the sky opens: there above me is the balloon. Clinging on to the sled, three lads. A hook is let down, and on the end of it, a thermos of coffee. To complete the feast: two croissants thrown down.

Wet, happy—I said to myself, the richest maharajah has not enjoyed a breakfast like this: a hot coffee brought by a dragon in multicolored livery to the summit of a giant lost in the forest.

From the first moment of our meeting in the Makande—it was dark—Francis said to me: you must come and see my garden tomorrow.

An irregular piece of land, shaded by shrubs, not far from the river. Among the chaos, a collection of plastic containers. In them are growing young plants: seedlings, cuttings taken from the nearby forest. The amateur would have seen it as the waste from a nursery, at best a sick bay. I see in it an approach to diversity, an attempt to understand some of its representatives. This work, liberated from the concessions made to the rules of garden art, encapsulates in all seriousness, wise to the point of madness, the quest of a planetary gardener, an ordinary man.

I haven't said that, while describing his plants—a delicate litany—Francis's eyes were shining, his face transported by an internal happiness, and everything about his expression reflected an imperious mission: to observe.

At first sight, scientific research appears to have no more than a remote connection with the garden or the gardener, the subject of this text. The work carried out by the treetop raft in the wider context of the planetary garden certainly has its place here: it participates in the exploitation of diversity without destroying it. One couldn't find a better way of defining the garden in its original concept: an enclosure intended to protect the best. The best of fruit and vegetables—nutritious flora, diversity exploited—the best of trees and flowers, the art of laying them out. Over the course of time it was precisely the layout that gained importance, to the point of becoming an art. The art of gardens expressed its excellence through architecture and ornament. These criteria are no longer sufficient. The life that develops in the garden, because it is threatened, becomes the central issue in such arrangements. Without prohibiting them, this responsibility erases the priorities of earlier times: manipulating perspective, laying out landscapes as paintings, composing clumps, organizing festivities and amusements, and so on. From now on, the living organisms must be our main preoccupation. To consider them, know them. To form a bond of friendship with them.

Observing could well be the right way of gardening tomorrow.

Discoveries are rare. But what remains to be understood about a world torn apart by unfeeling knowledge is immense. It is easier to describe and enumerate diversity than to understand its mechanisms. To do that, requires observation. The ethologists, scientists devoted to the study of animal behavior, the botanists of plant behavior (these have not been given a name), the ethnologists and even the sociologists all work in common with the fluctuating data concerning living organisms. At no point is it possible for them to describe a situation as stable or definitive. They have to be satisfied with the temporary position: a solution to life, valid at that moment. At that moment only. Those people have everything to offer us. Their teaching destroys certainties.

If people paid attention, not to the result of these observations, but rather to the philosophy that emerges from them, they would take an entirely different approach to the laying out of space.

Instead of making a rigid framework to the garden or public space, it could be seen as flexible and profound, capable of absorbing the transformations of living organisms. The layout of avenues, the design of steps and basins, would appear in a subtle way thanks to transformable materials—earth being one of them—in order to be able to adapt spontaneously to changing environmental conditions. Maintenance systems, furniture, instead of being installed forever, with the risk of being quickly outmoded, obsolete, or simply inadequate, could be discreet, transformable, or even done away with. The choice of species adapted to their surroundings constitutes the real work of the gardener. The choice of a technology capable of supporting any kind of species is comparable to the reaction of the consumer weighed down with advertising. The ordinary amateur, immersed in the obscurity of catalogues, remains unaware that in the garden there is much to be gained by working with nature and much to be lost by working against.

Achieving this is not simple. It is necessary to relax the rules at the same time. Sometimes to abolish them. To forbid the insurers, manufacturers of contentious issues, any access to the public space that they seem to have definitively taken hostage. The hardening of the outlines of the garden, the remorseless characteristic use of concrete, clean and grim, is not only due to the vain desire of designers to make their mark on the space (inflict their signature on all its users). It is also the result of a stack of health and safety rules delivered by the plateful to the stunned designer who can do nothing but carry them out: railings, edging, height of steps, vehicle width, turning circles, everything is normalized, obligatory, calibrated, indexed, systematized.

Who would have guessed that, at a time when towns are paralyzed by bottlenecks and pollution, the "users," instead of rushing for the small economical vehicles that exist on the market, would have chosen to drive obese and extravagant 4x4s?

The user, an unpredictable creature, does not respond to the linear logic of reason. He echoes his desires, or what he believes them to be. He has been told to buy a shiny red machine that will cross deserts with the family without crushing a fly. So he buys it. Rich France, society of self-gratification, the curbs have to be broken up, everything begun afresh, the turning circles recalculated, so that your Rangers, half basket, half tadpole, perched on tractor tires, can be parked alongside the garage that has become too small, after long having occupied the tarmac of the rue Rivoli, the Champs-Elysées,

Garden of the École Normale, Lyon,
Gilles Clément, 2006.

and other spaces where, it goes without saying, the cattle guard is indispensable.

When the object is hard, it is not remade. It is broken up. Another is manufactured.

The broken object goes to the scrapyard. It is a piece of scrap. One more to increase the huge pile that threatens to overwhelm humanity.

The fact that art, in its creative obsession, is beginning to exploit this scrap, demonstrates that it is not only seeking to question society about its irresponsibility, but also exploiting matter. For matter there is. Indeed, there is nothing else: a collection of products laid out as a landscape, the elements, having become refuse, decomposing or recomposing at will. The virgin landscape having disappeared from the civilized planet, what we are looking at is in every respect the result of a "secondarization" of natural space. This term, used to describe nonoriginal forests, the result of human exploitation, can be applied to the whole of the anthropomorphized territory. Europe, as a civilized continent, represents the secondary stage of a landscape that we can no longer envisage. One can only imagine a landscape dominated by living organisms in which human industry left no trace. Today we are in an evolutionary phase ih which manufacture, in the form of inert matter produced by industry, takes precedence over the living mass (biomass). Organic matter is recycled automatically. Inert matter produces scrap. Europe can be seen as a collection of scrap, potential or actual: architecture, transport networks, subterranean networks, aerial networks, consumer goods, and so on, piled up (towns, villages), interlinked (streets, paths, road systems), or inserted into "secondary nature" (cultivation, stock-raising, forests).

The same description could be applied to other continents, Africa or Australia, by adding a category that is impossible to find in Europe today, except in the case of relict landscapes: primary spaces, territories that have never been exploited.[27]

This is how the garden is designed: with more or less "encumbrances," that is to say, objects that resist natural decomposition and by this very fact, occupy the space permanently.

Architecture, when it is expressed physically, is an encumbrance to the garden.

Architecture, when it represents an idea, makes the garden sublime.

---

[27] Examples are a primary forest in Poland and the mythical Forêt de Derborence in Switzerland, described by Ramuz, protected from humans by its situation: a shelf separated from civilized space by vertical walls at the top of a mountain in Le Valais. The project "L'Ile Derborence" in the center of the Parc Henri Matisse at Lille was inspired by it. It symbolizes not a new primary forest—impossible to reproduce by definition—but an ideal forest of the future, the result of planetary intermingling.

The talent of the artist is revealed in the delicate equilibrium of this balance between encumbrance and sublimation. The treetop raft represents the minimum of encumbrance in return for the maximum of ideas. The mobile, light architecture of exploration and exploitation: balloons, rafts, gondolas, inflatables, cables, etc., nomad encampments. Varied projects: the inventory of diversity, the recognition of behavior, the search for new molecules destined for the pharmacopoeia, for cosmetics, for the general improvement of living conditions. Nature, treated as a garden, as the object of research, does not generate any scrap. The displacement of objectives at the heart of this garden does not produce any disruption of the territory, any detectable traumatism, any hard landscaping. It is in its exploitation that everything begins. In order to produce a medicine developed from discoveries made in the tree canopy, it will be necessary to construct and lay out in the landscape the following scrap:

- a laboratory for the research center
- a factory in an industrial zone
- a commercial distribution network
- a conditioning process, packaging
- a sorting center for the packages
- an incinerating plant for nonrecyclable waste
- a center for exploiting recyclable waste
- a hospital.

From the artistic point of view, it is a failed project: it occupies a disproportionate amount of space in relation to the service it offers.

In an ordinary garden, a vegetable garden let's say, the total number of encumbrances occupy less than a tenth of the territory: shed, tools, fencing, edging, planking, greenhouse, frames. Nevertheless, the garden yields a considerable volume of products: fruit, vegetables, roots, seeds, flowers, wood, all organic matter capable of being transformed into refuse. Recyclable. Immediately.

Hence, the compost—principal area, heart of the garden. Everything ends up there, Everything begins again from there. The mysterious veils of aesthetics and ethics having been drawn over kitchen peelings, compost has been relegated to the back of the garden when it should be at its center. Organic refuse is transformed into food. In the predatory chain, it always ends up on the human table. That is why, in the long run, it is so important. The future of humanity depends directly on its way of life. The richest country on earth is clearly in the process of committing suicide. The planet finds nothing wrong in that, perhaps it would even be relieved. This McDonald-sized portion of the land above water level carries with it the majority of civilizations as slaves to its system. Those who subscribe to alternative globalization—since that is the term—manifest their intention not to die immediately, they have caught sight of another way of gardening. One in which the question of refuse, central as it must be, is not treated lightly. One in which vegetable peelings—an economic argument—are raised to the rank of energy resources instead of being consigned to the trashcan. As in a garden.

Gardening lends itself to a management system extending beyond the bounds of the garden. It is easy to understand why a society takes inspiration from it—it's the main issue of *Le Jardin planétaire*—provided it perceives certain aberrations and avoids them. For example, what is the point of the mechanical blower, a noisy tool, polluting, evil smelling, throwing up dust and humus, all nutritious matter, in order to pile up with great difficulty a few leaves from the ground? To make it clean. The death of a clump of trees in the Parc de la Tête d'Or in Lyon was caused by ten years of blowing, depriving the trees of humidity and nourishment. A ridiculous seizure of power, aimed at maintaining a space according to an ideology. The defoliation of the mangroves with napalm shares some of the characteristics of these garden aberrations, as well as all the wars, from Vietnam to Iraq, including the treatment of corn and of poor Madame Meilland (Peace) roses that accumulate far too many aphids.

The rule and the idea, magnificent servants, are transformed into absurd tyrants as soon as ideology and regulations appear. Having

come to visit the great kauri[28] in the Blue River Park, a nature reserve located in the south of Noumea in New Caledonia, we were hoping to see the strange creature that inhabits these regions: the cagou. A bird with red eyes that does not fly but barks is worth a detour. On the lookout in the Caledonian jungle, we were hoping for a yelp, a wail, any kind of animal noise reminiscent of a dog, to locate the cagou among the thick vegetation. Instead of that, a whirring sound from beyond the boardwalk on which we were walking. A man in charge of a blower was undertaking the cleaning of the undergrowth; the dead leaves had to be removed. In a dense natural tropical forest there is no lack of leaves. The task was huge. Luckily, we were told, it only applied to the wooden boardwalk; someone might slip, and then . . .

A broom would have been enough, the man in charge of the blower agreed, but the pathway was designed in such a way that sweeping it was judged to be impossible. Why?

Admirable civilization, just listen to this recipe of utter crassness: take a primary forest, unique in the world, and raze it to the

[28] *Kauri: (Agathis sp.)* one of the great conifers of the Southern Hemisphere.

ground. After exploiting all the fine kauris, leave just one, saying it is sacred. Immediately, create a reserve to attract the ecotourist eager to see cagous and native relics. Design a pathway, neither too long nor too short. Build a boardwalk with the aim of making it convenient and usable in all seasons. Then, immediately, throw the book of health and safety rules at it: the raised pathway represents a danger, it must have edges to prevent wheelchair accidents. Construct the edges.

Let the leaves fall.

Realize that because of the edging, it is not possible to sweep without dragging the heap hundreds of meters.

Blow.

At which point, the highly protected reserve, armed with previously silent equipment—duckboard, trashcans, labels, benches—is to be enhanced by noises that were not there. Motor engines, intentional tornadoes, you have all the rights, including that of transforming nature into an experimental territory for stupidity.

Martial, a Noumean geographer, accompanied us. His strength enabled him to manipulate his wheelchair alone. He went everywhere, transforming his handicap into part of his personal gear, the same as the photographic equipment he used. In his opinion, an extra twenty centimeters on the sides would have made it possible to avoid the dreaded edging. Impossible, replied the ecoguide, pretending to look for the cagou among the thick foliage, nature's paths have precise measurements.

Ah.

No sign of a cagou. On the other hand, a lot of blowing. And three young boar on the spree, following in line, as pigs do all over the world.

Having turned back, after visiting the great kauri and going into raptures over it, taken a few photos of the limpid Blue River, the guide reassures us: the cagous are not far off, there is a way of attracting them. He takes a cassette out of a pocket and inserts it into the player of the 4x4 through the open door. "Listen," he says.

Admirable civilization, listen closely to the song of a rare bird, emerging from a piece of scrap metal in the depths of a voiceless forest. That is all that is left of an ancient music. But the technology is there to save us and to protect our memory.

In the Ecomuseum of the Blue River there will be:

- a cloned kauri tree
- a waxwork ecoguide
- a biscuit-colored 4x4 pickup
- a pair of binoculars, worn but reliable
- an audiocassette and its antique player
- and in the self-serve cafeteria, a choice of Big Macs made the way grandmother made them at the time of the Kanak[29] civilization.

---

[29] The Kanaks are the indigenous Melanesian people of New Caledonia, annexed to France in 1853.

On the blackboard, clearly framed with bold strokes, hatched with stumps of chalk, one short sentence like those in old grammar books: subject, verb, complement . . .

It sits on the worn surface of the blackboard, now cloudy like the objects around it: the brush, the back of the chair, the floor of the lectern. Someone has been here in the night. Yesterday, there was nothing yet on the walls to support or to break up the resistance that we were trying to mount. We: a group of neglected[30] students, inspired by the spirit of revolt and the desire to construct a new space, a pedagogy and a curriculum worthy of our ideas. Dreaming aloud.

Through the democratic process, I find myself "president of the committee for reform," an invented title that I hold for a few hours; the discussions that follow, spent deciphering the mysteries of the little sentence, cause me to stand down. For the moment, I don't know what to think. It is not some kind of official announcement, administrative or technical, seen every day hastily pinned up on any kind of post. Nor is it something left over from a required seminar, a comet tail, the final message before the bell. The height of the letters, the placing, everything is presented as a signal, with its impact and its economy: a revelation.

Once read, it expands, fills the space of the lecture hall with its lines of empty tables, reaches the furthest corners of the hall, the box that houses the silent projector: makes its way outdoors, through the windows, the fanlights, any openings. Everything is too small to contain the words that I read and reread, in the rhythm of a textbook:

"The tree is a capitalist."

Can the sentence be turned around? Capitalism is a tree, its roots anchored in a compost of hidden factories, its branches exposed to exhibit the best products. Where does capital hide in the shady architecture of the oaks? Is it possible to incorporate vegetable

---

[30] *En déprise:* a phrase applied in French to fields where cultivation has been abandoned, and also to elderly people deprived of their identity. —Translator's note.

synthesis in a system that permits the accumulation of riches? Could the annual growth of trees be a product of this accumulation? What profit could the tree derive from such a mechanism, beyond ensuring its own life?

Instead of a metaphor, should it be seen as a symbol? The tree, as a prototype of urban charm, a complex hygienist, a Haussmannian ornament, a bourgeois amusement. An emblem of "green spaces" accompanying the whitewash of grass at the foot of the HLM (public housing). A degrading model of a social contract whereby nature is distributed like anonymous alms: 1 percent considered obligatory. Was it related to this offense: we are giving you a bit of greenery, so keep quiet?

The meeting that followed shed no light, light came from outside along with the excesses of spring, with the sun so high, the vigor of the grass and the bodies of an impassioned crowd. A mysterious light in the suburbs strewn with black flags where processions advanced in clouds, shields, shouts, teargas, where Paris gave way to the creative impulses of the spirit, where the very paving stones generated slogans.

Certainly, there were discussions, ideas, but the urgency seemed to lie elsewhere and Versailles so far away—once again, too far away—receiving messages on the wind, like spray. I terminated my studies there. The training of a landscape designer is still dependent on a course at the École Nationale d'Horticulture, said to be obligatory for learning to approach gardens appropriately. Tiny illuminated fragments from the capital reach the sleeping towns around, either by chance or misfortune. On the blackboard, in large letters, one of these critical fragments determined our actions: an invitation to engage our cautious reforms with the real revolution. There were no more than six or eight of us involved in the debate. One of us ventured the hypothesis that a couple had come specifically from the Odéon, where the revolt was in full swing, to reveal to the ignorant inhabitants of Versailles the truth about trees and humans. It was time to get out of here.

My mother was growing bored with the Right, where she never-theless had certain attachments and convictions. She rode to the barricades on her Solex moped. The light of the riots reached as far as rue Claude-Bernard where we were living. Something was finally happening. My father, being anxious by nature, was building up reserves of petrol. As for me, I decided to collect plants.

At that time, the National Route 20, passing through the Berry region of Champagne, was like a country road. On the side of the road around Massay it was possible to park without obstructing the traffic. There I found three sorts of bee orchid (*Ophrys*) together with another delicate orchid: the man orchid (*Aceras anthropophora*). Since then, I have never again seen this little flower, with its extended labella, found on chalky soil and nowhere else. I made my way to the Creuse, a soil so acid that it is covered with little sorrel in the spring. The orchid family is not well represented there. Walking through an abandoned field, that same day I came across a tiny burnt orchid (*Orchis ustulata*), accompanied by blue Polygala and later, in the ruins of the chateau of Crozant, a Loroglossum (*Loro-glossum hircinum*), smelling of billygoat, growing out of the fallen rocks. None of these plants should have been there among the gran-ite outcrops. The lime mortar and fertilizers might explain their presence.

The majority of species that make up my herbarium date from the spring of 1968. There was everything I needed: space, time, and good weather. Rather than discussions about liberty, interminable and heated, producing nothing but a web of hypotheses, I prefer experience. While waiting for society to get round to establishing the civic laws that it intends to impose on the people, it always seems possible for me to embark on a journey. The noise of the skir-mishes reached me in snatches, as if from afar, like the frolics of a game to which I was no longer invited. Can traveling ward off the feeling of rejection that we sometimes experience? I wanted to see. That was how I learned to let my eyes take over.

From this familiarity with things observed came a sense of solitude, about which I can only say that it has brought me closer to what we generally call the landscape.

It wasn't written in books. There was nothing to predict this wandering. I have never returned to the starting point. The route adopted that day has never ceased to lead me round the world. Each time, it follows a new course. Geography, changing light, people met on the road, different looks inevitably exchanged.

The little sentence on the blackboard keeps it secret. Laid to rest, lurking in who knows what kind of secret shelter, it is working in the shadows. Sometimes it emerges, hovering above legitimate texts, hermetic and violent, on the point of conveying a message, without ever succeeding.

For a long time, I refused to accept the improbable marriage between nature and politics. No less, the conflict opposing one against the other. An impossible meeting on fields of battle so far apart that the combatants—or allies—would have no chance of confronting one another. Besides, I didn't use my right to vote. What should I throw into the urn, whether opaque or transparent? Papers scribbled with untenable promises, endless speeches, faceless and meaningless. The Republic gave me the right; I abstained.

Sticking to gardens, to the endless landscapes from which they are made. That was enough.

The political question remained unresolved, or rather silent, no doubt waiting for the conditions that would make it possible to emerge.

The opportunity arose in the form of an awakening. Suddenly, the speech of one man became comprehensible. It was my first vote—truly. Admittedly, it was for a man, not a politician. Never mind, I voted, adding my voice to those few who had come to support the candidacy of René Dumont. The first person to take up arms on behalf of ecology, at a time when that term, discredited today, still had a meaning, you may perhaps remember. He was holding in his hand an insignificant glass of water, tiny but fit to drink. The time will inevitably come when we talk about a period when it was possible to drink from springs.

The result of those elections, barely 1 percent, strengthened the sarcasm of the conservatives, held up to ridicule the champions of life. It filled France with a shame from which it has never really recovered. As I write, the minister of ecology bathes in the oil with which he himself feeds the marshes, swims delightedly in nitrate, fishes, hunts, pursues old habits to his heart's content. As soon as he is wounded, he cauterizes himself with ecomuseums, and to fill his spare time, classifies a pretty landscape here and there, destined for the tourist guides.

Thirty years separate May 1968 from May 1998. On that date, I drafted the scenario for the exhibition "Le Jardin Planétaire." Commissioned by the management of the Great Hall and the Park of La Villette, it came a few months after the publication of "Thomas le Voyageur," an essay providing the basis for the discussion. I had chosen to speak about "ecology" without using the word itself, distrusted on account of so many battles, so much hesitation and radicalization. "Garden"—capable of rallying the public on common ground—seemed more appropriate. This term, when associated with the planet, extends the horizon of the ordinary garden, opening it up, like any globalization process, to a limitless citizenship.

The way in which one conceives the world has a direct bearing on the way one treats it. "Gardening," as an expression of cultural diversity, threatens or protects natural diversity according to the methods it employs.

The extension of the garden to the scale of the planet proposes that all agricultural techniques come under planetary gardening. For a specific crop, take rice for example—the basic food for a significant part of humanity—the protocols vary across the planet, in spite of a certain standardization due to globalization. What difference is there between Bali, a Hindu fragment of Muslim Indonesia, and the Camargue, or Christian Louisiana?

In Bali, when the rice is ready to be harvested, offerings are made to Dewi Sri, the goddess of rice. Two figurines, woven out of palm leaves, fixed separately on the thatched roofs, represent male and female, a symbol of fertility. The birds and rats make off with part of the grain. An arsenal of scarecrows and windchimes are added to the shouts of the guardians of the rice fields. They drive away the excess sparrows—*burung padi*—whose delicate bodies come and perch on the ears of rice without breaking anything. The way of frightening the birds varies in different sectors of the island. The human silhouettes dressed in cheap finery that you see in our country rarely appear among the crops. The Balinese prefer to use shiny pieces of material stretched in a web held by invisible threads. Thin strips of transparent plastic, cut out of grocery bags, moving in the slightest breeze, mark out the rice fields in this way. To anyone unfamiliar with it, the general scene—music, banners—looks more like the work of an artist than the labor of the fields.

At the foot of the figurines, they lay cakes of pink rice, frangipane, and hibiscus flowers, a few grains of white rice and fruit. The person officiating, dressed in a new sarong with a wide yellow belt, throws on the temporary altar drops of sacred water that comes from Mount Agung, territory of the gods, blessed at Besakih, the mother temple, and distributed throughout the island. Finally, they sing the appropriate mantras in an ancient *kawi*,[31] tracing coils of smoke with sticks of incense.

In Louisiana, when the rice is set, a final spray of pesticide is delivered by airplane. The Bible predicted no less.

In one case, you will please the gods, the rice, and the creatures that populate the universe. In the other, you will make money.[32]

---

[31] *Kawi:* a poetic and sacred form of Sanskrit used by priests.

[32] Nowadays, the central Indonesian government obliges the Balinese to use a new kind of rice, producing three harvests instead of two. This variety, *Orize sativa*, prone to disease, requires treatments that threaten the ecosystem of the rice fields. Dragonflies and their larvae, frogs, snakes, eels, fish—all sources of nourishment with extra proteins—are tending to disappear.

An entire management approach depends on the foundation of our beliefs. A whole political system follows from this method of management. And not the reverse. When by chance, at the request of some flourishing utopia or technocratic plan—consolidation, partition, and other "roadmaps"—a management method is derived from a political idea (this does occur), it is not with a view to discussing agricultural proposals, but rather to reinforcing them by simplifying them. The common agricultural policy is a tragicomic example of western Taylorism.

Nature, held hostage behind the visionary screen particular to each culture, pays the heaviest toll where the cultural system concerned places Man as master of the cosmos, and a correspondingly lighter one where he is placed on an equal footing with the other beings. The Indian animist of tropical America regards the tree or the puma as a human being that has taken the form of a tree or puma. When we look at a tree—or a puma, which is less often—the idea of an equivalence does not so much as enter our minds.

The feeling of being part of the world cannot coexist with the desire to dominate it. In fact, the animal sacrificed in the name of a belief, the one killed for food, or the one hunted for pleasure, share the same oblivion: they disappear from the surface of the globe. But in any of these cases, the relationships of Man to his environment, to this animal, for example, are not the result of similar visions of the world, of similar distances from, or of similar promiscuities played out with the beings around. To live according to the convictions that everyone believes in: that is the subjective element of space. That is why the "landscape"—the terrain affected—contains so much that is both spiritual and material at the same time. Whereas the "environment," spanning cultures with its objective and universal vision, drains space of its content and saps beliefs, until nothing is left on earth but a heap of cinders called biomass.

That is no reason to abandon the measurement of the world and the tools of science. Ecology, through its special dimension, a sort of "biometry," sheds light on previously unknown facts about our universe. The idea of a planetary garden would never have seen the light of day without seriously taking into account ecological parameters. The very term "garden," meaning an enclosure (from the German *garten*), could not be applied to the planet without retaining its meaning. The emergence of ecology, an event without precedent in the historic relationship of Man to Nature, coincides with a systemic and hence global vision of living organisms (ecosystems). At the same time, it fixes the field of exploration firmly in the very heart of the biosphere. A terrifying revelation: the earth, understood as a territory reserved for life, is a closed space, limited by the limits of living systems (the biosphere). It is a garden. Once said, this statement obliges all human beings, as passengers on earth, to shoulder our responsibility to protect the living organisms whose steward we are. So we are gardeners. A role for which society on the whole—apart from a few indomitable animists—has not been prepared. This explains the considerable difficulties experienced by the majority of populations in trying to reconcile their beliefs with the brutal attack on their conscience provoked by the ecological idea. This also explains the excuses, the various twists and turns, the diplomatic, economic, and political maneuvers undertaken by the powerful states, to the point of exacerbating all kinds of fanaticism, in order to justify any approach allowing them to escape the grip of ecology. The reconciliation of beliefs and interests with the requirements of "gardening" will take a long time and perhaps will never happen.

While waiting for what perhaps will never happen, I have learned to see as a whole the objective attributes of the place in which we find ourselves, and those that are indefinable, that people call subjective. Together, they define the terrain over the course of centuries. Every fragment of space that has been anthropomorphized may be considered as a palimpsest on which the great visions of the world are engraved, one over another. From now on, we shall have to add a superior state of awareness, defined not only by the interaction of living beings, but also by their cultural systems: an eco-ethno-system, at the same time both unified and diverse. A vast garden, a small planet.

So I shall never know whether the tree is a capitalist, or whether, in opposition, the Marxist grass unfolds at its feet like a rebellious and trampled crowd. I have not investigated the Pétainist tendencies of sainfoin, the Nietzschean aspect of the saxifrage in the depths of the woods on the banks of streams. But I do know that all living things, whose sole aim is to live, will one day or another be called upon to become a political object—whether formidable or minor.

The subject of the discussion.

It concerns their very existence.

CONCEPTS, "CONCEPTEURS"

The encounter with an unknown site adds an image to the collection of landscapes stored in the memory. A sensitive picture. Open to interpretation.

This phase involves an emotional reaction without which it would be impossible to hope for a point of view that one day will inspire a project. One can only speak correctly about a site after going there. Written documents, illustrations—photos, drawings, maps—films, reports, can never replace the total immersion that produces the first impression. One should not underestimate this time, the project is contained within it. Sealed. Implanted in the air, elusive for the moment, but present. The result of a confrontation between the environment, as a complex transmitter, and the person, a complex recipient.

The gathering of objective data corresponds to a second phase. Any site anywhere on the planet has its own coordinates: whether insignificant or detailed, simple guidelines or hefty documents. Certain ones have the benefit of extravagant environmental impact and feasibility studies. Their aim: to obtain the agreement of those inhabitants, whose way of life is about to be destroyed. Sometimes, their very lives. Displacement of populations, flooding of valleys, construction of motorways: the planning of public amenities permitting no discussion. Impact studies do at least offer the advantage of bringing together multiple data: history, geography, climatology, sociology, and other knowledge available to impact experts.

Beleaguered, manacled, bombarded with information, on its knees, the site begs for a project. The "concepteurs" are ready.[33]

Concepteurs.

Custom demands that the architect of the project bears this title. Thanks to this turn of language, he or she becomes responsible for anything, whether near or far, that has the potential to cause a problem. Being a concepteur means shouldering responsibility for a

[33] *Concepteurs:* the word, normally translated as "designer" in English, is used in French to denote more specifically the person who generates and carries out an idea (or concept). I have therefore chosen to retain the French *concepteur*, particularly in the context of the text that follows.—Translator's note.

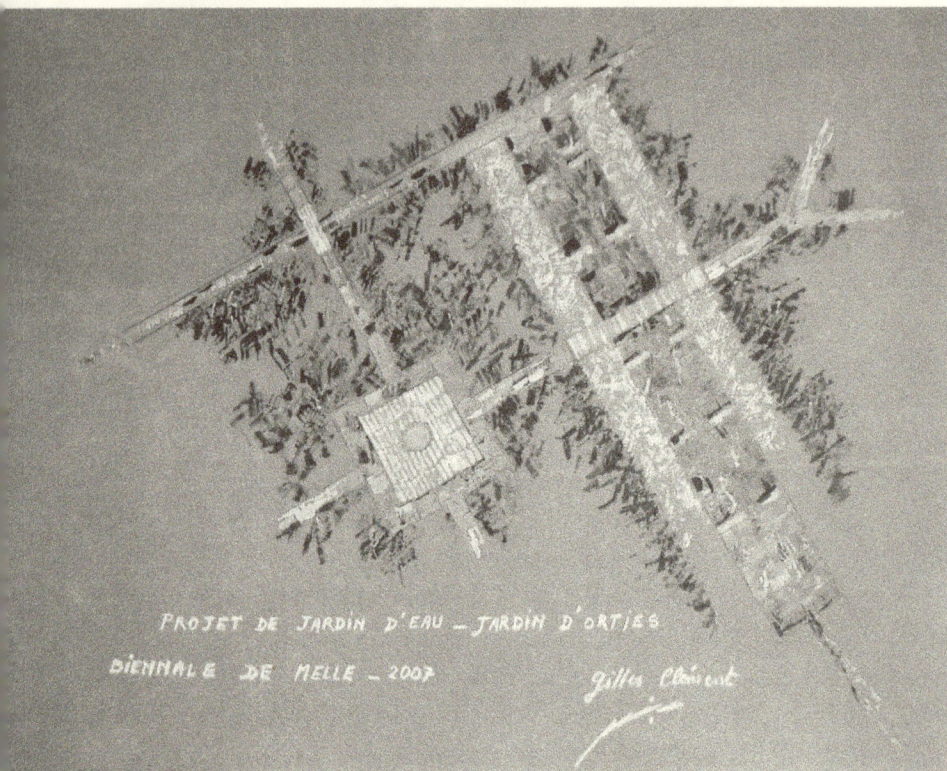

PROJET DE JARDIN D'EAU — JARDIN D'ORTIES
BIENNALE DE MELLE — 2007

Gilles Clément

layout intended for public use. At no point is there any question of a concept. The burden of contention, implied by the very title concepteur, relates to the very real imminence of any danger, which will demand legislation. The philosophy of the project is of no interest to the administration. They only want to know to whom they should turn in the case of litigation.

I was not so naive as to imagine anything different, when the superintendent of works for the Quai de Javel approached our little group, in our role as concepteurs. The Parc André Citroën originated from the amalgamation of two teams,[34] in which each of us shared the

---

[34] Patrick Berger, Jean-Paul Viguier: architects; Gilles Clément, Alain Provost: landscape architects.

same dubious honor. Until that day, I had been included among the landscape designers, or gardeners, or gardener-landscape designers. I came to be considered as "architecte-paysagiste." Neither insult, nor compliment, just a literal translation of the English compound word, used in France to enhance one's prestige: landscape architect.

More often than not I am called by my name.

In France, major projects demand concepteurs. Everybody seems used to this language. The first time, terrified, I wondered where to turn to discover the concepteur in me. As the days went by, by examining the term from every aspect of my torment, I finally exhausted it. This type of word resists questioning. It can be admired, but then returns to the stratosphere of abstractions.

To escape drowning, one can consult the dictionary. On any occasion, apart from explaining the murkier literary areas, the book of books offers a solid lifebuoy. A few stubborn words resist illumination. When it comes to the word "concept," the grammarians seem to have given up: "In philosophy, idea, object conceived by the mind," adding: "French scholars have created the word *concept* to translate the famous *Begriff* of Kantian philosophy, and it can be applied to any general notion without being absolute." With an irritated stroke of the pen, the editor makes it clear that one may well wonder whether it was really necessary to expand an already rich vocabulary: in his opinion, "notion," "idea," "object" suffice. The nineteenth-century Grand Larousse ignores the word *concepteur*, described as a "little used neologism" by its contemporary, the Grand Dictionnaire Universel.

From *concipere*, to conceive, all the words derived from this Latin root evoke "creation," in the sense of the reproduction of a being: immaculate conception. A little Pléiade of words follow in alphabetical order. "Conceptual faculty" is mentioned, but also "conceptualist": the name given in Spain to those exaggerated cultists, poets who are only prepared to accept forms not in common use.

However, the developers—our clients and sponsors—somewhat distanced from Kantian philosophy, stick to a simplified vision of the word that is not mentioned in any dictionary: the concepteur is the one who draws the project and names it. He is responsible for an image produced by his overheated brain. Draftsman? No, the draftsman is to the concepteur what the gardener is to the landscape designer: the one who performs the task. Important not to confuse things!

Having arrived at an unknown site, is the landscape designer going to respond to his supreme duty: to produce a concept? Will he have enough distance to detach himself from the emotions that bind him to these places? To produce a line of argument? Establish a theory, define a vocabulary, translate his discoveries into realities? Show astonishment that no one before him has grasped the importance of the obvious features that he has been able to list? Although so easy. Is he on the other hand going to give in to his instincts, draw, shout, laugh? Make a little installation?

In reality, the concepteur, only human like anyone else, bends beneath the files of "impacts," deciphers the local constraints, tries to unravel the contradictory interests of a multiheaded client, dissects the programs, if necessary inventing them, sets up a team, and throws himself into the combat of negotiations with the technical bureaucracy.

Crushed, caught between bills to be paid and the demands of the bank, the concepteur ends up by scribbling a design, a sort of confession extorted under the torture of contingencies. This, he says, is the best I can do, it's an idea, it's good, I am defending it. And he invents a statement that explains the discrepancies in the drawing. A carefully gauged statement, sharp, and if possible, "conceptual." Yet clear and sprinkled with acceptable numbers. A statement submitted to the administrative hydra that chooses from among others the one that it reckons will be amendable.

As soon as the winner receives the verdict the ordeal begins, the concepteur, summoned to the keyboard, has to fill the memory of hard disks before giving birth to an enormous consultation dossier—like a flabby marionette without bones.

The studio of a concepteur consists of computer screens, shelves of catalogues, walls of files and drawing boards on which no one draws. His time is divided between meetings: over the phone, online, or in person, each time a report is prepared. Duplicated according to the number of participants, the report is added to the mass of files, not a leaf of which must disappear in the course of the next ten years.

Sometimes, rarely, the concepteur is lucky enough to see one of his projects realized; if he retains control over the site, he can hope to be delighted with the result. Or dismayed. The rest of the time he creates paperwork.

With each project a forest disappears.

Of all the concepteurs, the landscape designer is the one who can hope one day to replant the forest that his harvest of projects will have caused to disappear. An advantage.

The role of gardener rarely involves a concept. At least, that is what they say. The gardener gardens. The gardener does not do paperwork. Or at least, very little. The gardener's hands are occupied and his mind free.

To escape the tyranny of the limitations imposed by society, I call myself a gardener. My mind is not as free as I would like, but my hands are occupied. Before the house was built, I had begun the garden. The two have grown together in the space of a few years. I went from the mason's trowel to the rake. The house, an inert object, comes to an end. Whereas the garden never ceases to begin.

No longer using the trowel, except when absolutely necessary, I persist with the rake. Sometimes with the pen. That was how "Le Jardin en Mouvement" (The Garden in Movement) was conceived and published in the spring of 1991.

The emergence of a "concept"? Nature, revisited, becomes a principal actor. The gardener invites her to share in decisions. The garden returns to us in the guise of images that are sometimes familiar—*friches*, fragments of nature, flowering fields—but the principles on which it is based, along with the way in which it is managed, have no connection with past models.

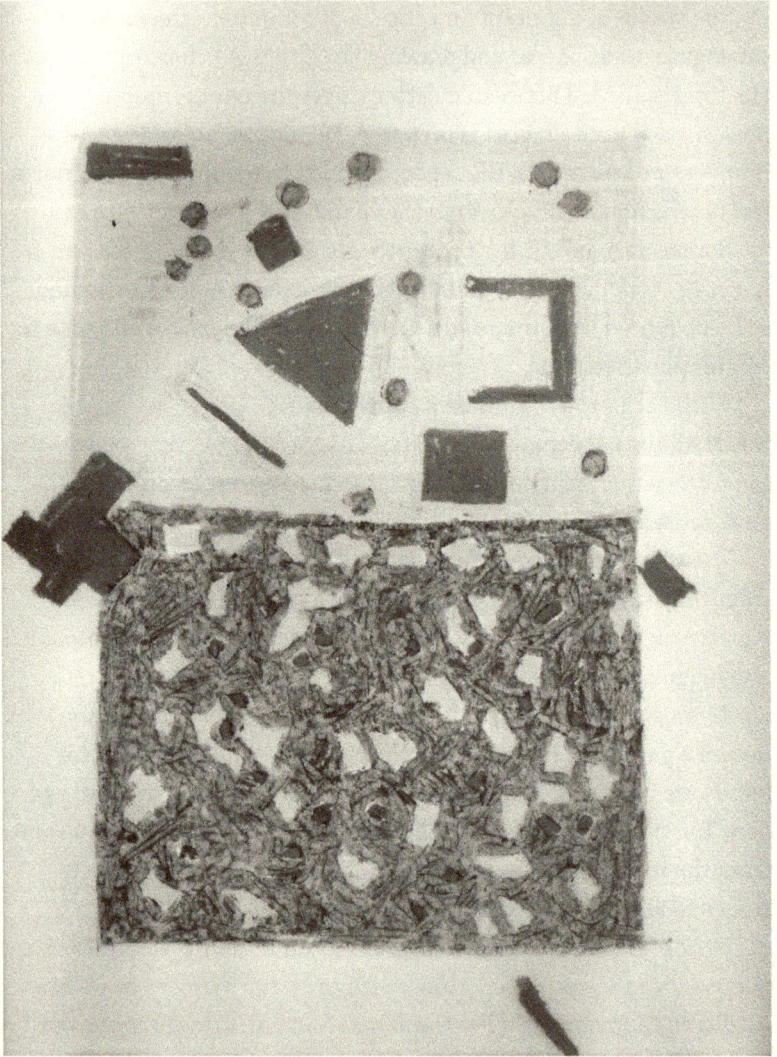

A conceptual image for a three-hectare park surrounding the
Museum of Concrete Art in the town of Mouans-Sartoux. It is based
on two principles: the geometric forms in the top half represent buildings, museum, chateau, in a clearing; the bottom half of the image
represents a wood broken up with particles of light. Gilles Clément.

The reality lies entirely in experimentation.

Uniquely.

Without gardening, the garden doesn't exist.

You can garden, respecting tradition. You can garden differently. From the moment I bought a plot of land in 1977, I had chosen this route: to discover nature before controlling her.

By paying attention to the behavior of species, I had become aware of the conflict between their natural inclination to unfurl, reproduce, rest, and our desire to "beautify." Impossible to reconcile these two positions, as long as one remains a slave to the aesthetic canons of "garden art." How to find a way of making the biological reality compatible with an aesthetic?

If one admits that biological reality directs gardening, there is only one thing to do: make it subservient to its own aesthetic, declaring it to be valid because it's right. Could one admire the nettles, wonder at a *friche*, discover in the richness of abandoned territories the material from which to make a different kind of garden? Accept the appearance and disappearance of species in places where unchanging forms are expected? Devoted to this task, I ended up by admitting that movement alone—understood both physically and biologically—made it possible to resolve a difficult question: how to welcome living organisms in such a way that the changing forms, at which they excel, never throw the gardener into confusion.

That meant accepting the garden as a terrain entitled to changes of aspect, colors and even flowering successions. A threat to architecture, seen as the only way of understanding space. A new aesthetic was to emerge from this position, based on principles drawn from a model offered by scientists rather than plastic artists.

By stating the principle of equivalence between layout, as we imagine it in its geometric aspects linked to inert matter, and information, seen as the formal transmission of biological messages, Henri Laborit suddenly reveals a way of reading nature that makes any kind of abundance, judged elsewhere to be chaotic or inelegant, acceptable because it is comprehensible. When applied to the garden, this vision of the world enlarges the field of aesthetic and formal tolerances. It allows us to appreciate a moment of abundance in nature—no matter how it reads formally—content to enjoy the spectacle, not because it is beautiful but because it is understood. In order to develop the Garden in Movement (a section of the Parc André-Citroën, where plants are allowed to re-seed themselves naturally, albeit ultimately controlled once a year by the gardeners, to prevent one species from taking over), I initially adopted this position. Later I realized that it was possible to combine a formal aesthetic—which we are accustomed to in gardens and landscapes—with the equilibrium of an ecosystem.

The theoretical part of the discussion should be seen as the analysis of a focused experiment, that is to say, research. Exploiting diversity without destroying it requires the implementation of living organisms and the presence of the gardener (the site manager) to verify, inflect, or oppose their supposed capacities for movement. It requires a plot of land, in the literal sense, a surface covered with earth, a soil.

In making a garden, I did not intend to convert practice into theory and turn it into a book. It was enough to struggle physically with the garden without writing about it. For a long time I gardened without clarifying my ideas. Nevertheless, there were plenty of standpoints: to conserve the diversity already present, to increase it, to utilize the energy inherent in the species, not to use opposing energies unnecessarily, and to end up with a pledge that I repeat as often as necessary: to do as much as possible with, and as little as possible against.

If the garden in movement, as a principle, is derived exclusively from experimentation, the planetary garden is the result of nomadic observation comparable to experimentation (traveling), combined with a hypothesis: can the Earth be considered as a single garden? Can the precepts of the garden in movement be applied to it? The food (or riches) available are endlessly diminishing for an ever growing number of consumers, while increasing for an ever diminishing number of privileged people. In parallel, the diversity responsible for generating all riches is being exhausted by expanding activities and sheer numbers. Impossible to imagine what might organize a balance between the predator and his prey, if not whichever of the two develops a conscience. Provided, of course, that it is shared across the planet. There is no planetary gardener, just as there is no localized example of a planetary garden. The planetary garden is a principle, its gardener the whole of humanity.

The proposal is to consider diversity as the guarantee of a future for humanity. It has to be recognized, documented, and protected. An infinitesimal part of this diversity, exploited over vast areas, contributes, alongside other lethal pressures—pollution, overpopulation—to the progressive but increasingly rapid reduction in global diversity. The important question posed by "The Planetary Garden" can be expressed in this way: is it possible to exploit diversity—having thoroughly recorded and understood it—without destroying it?[35] To go further: can the recording, the understanding of the mechanisms connecting these living beings with one another, but also the exploitation of all or part of these components, be considered as a means of saving diversity? There is no lack of examples of balanced management in nature. There is the famous case of the thorny acacias of tropical Africa, whose leaves produce a toxin intended to discourage impalas. Subjected to excessive predation, the plant not only produces a poison, it warns those of the same species through a chemical message that has long remained a mystery. A puff of ethylene received as a warning: antelopes are around, prepare your poisons. In the forests of Gabon, a shrub[36] harbors enormous and dangerous ants. They dig tunnels that damage the plant, which has turned this to account: at the least sign of predators—browsers are plentiful—the ants emerge and attack the attackers, who flee at the

[35] Such a question would not have arisen without a modern consensus: humans are everywhere, and everywhere they attempt to exploit everything. This anxiety dates from less than two centuries ago (cf. Lamarck, *Passeurs*).

[36] *Barteria fistulosa* (Passiflora family): Grows quite simply less well in the forest of Western Africa without the presence of ants (information provided by Francis Hallé).

first bite. It is rare for a plant to remain isolated from the predatory system (in the widest sense); either it serves as food, or it feeds directly on waste products (saprophytes), or it acts as a parasite on other creatures. It is amazing to realize that, in spite of the very varied ways in which nature exploits itself, species do not disappear, at least not for those reasons. On the contrary, certain ones owe their existence to the relationships maintained with the menacing creatures of their environment that serve at the same time as their protectors. It would not occur to an animal to exhaust definitively the food it needs every day. Its life depends on it. The nomadic behavior of herbivores is directly linked to this type of management.

Examples of the survival of species due to the relationships linking man with nature are much more rare. A considerable number of cultivated plants no longer exist in the wild. Certain ones, the result of progressive improvements, have never existed in the form that we know them today. Others have not changed. The difference between the wild apple *Malus sylvestris* and our dessert apple is a matter of size and flavor. The appearance remains the same. The cultivated varieties have been added to the botanical strains. In one sense, diversity has been increased through the number of varieties maintained artificially alongside the original species. We might congratulate ourselves on such an increase, were it not for the fact that in order to cultivate a finally limited collection of commercial species, we have ended up destroying all the others. An orchard of commercial apple trees contains apple trees and nothing else. Cleansed of any possible competitor, the surface of the soil reveals nothing but churned up earth, slimy and almost lifeless.

It is not the principle of exploiting diversity that is in question, but the method of exploitation. How can we replace the brutality of techniques said to be modern with a form of management that is sensitive, diversified, and truly modern? Adapt technical knowledge to the demands of planetary ecology and not, as still happens in all the rich countries, transform it into a servant for the sole benefit of pressure groups.

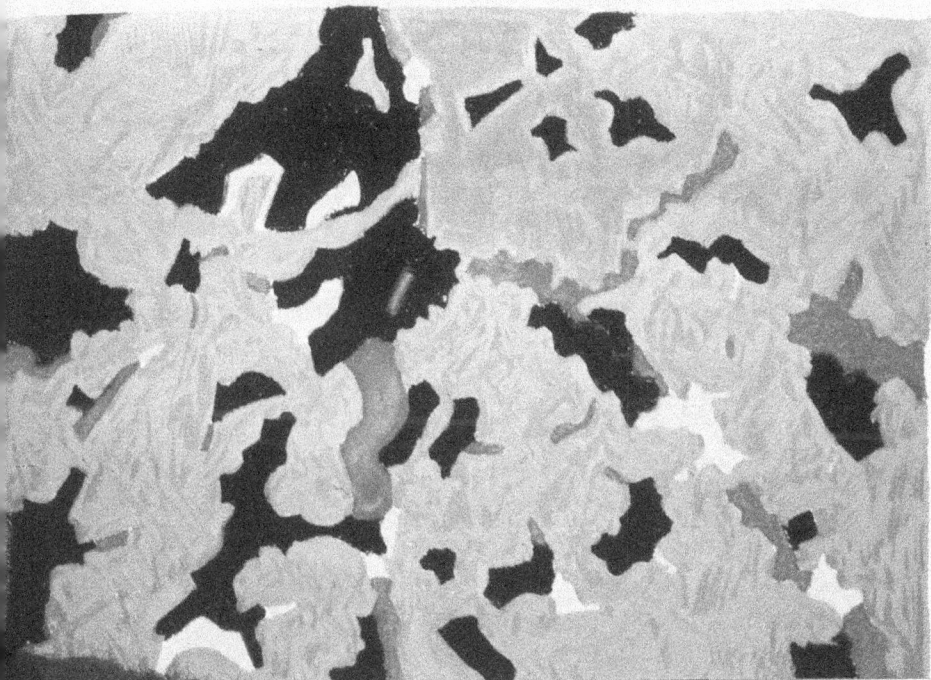

A preliminary sketch for a tapestry commissioned by the towns of Aubusson and Felletin (Creuse), depicting the third landscape in the valley of the Creuse between Felletin and Aubusson. The dark areas represent forest; the lightest areas human activity; the intermediate shade the "third landscape" on the slopes of the valley.

Yet even if it is desirable to understand the extent and originality of diversity—if only to conserve it—perhaps it is not necessary to exploit it in its entirety. One can even imagine a protected area where diversity would not be exploited in any way. Perhaps not even observed.

In 2002 the Art Center of Vassivière, now the Center of Art and Landscape under the direction of Guy Tortosa, commissioned from me a photographic study of the region. The result of the study can be summed up in three terms: shade, light, third landscape.[37] The eroded relief, ripples descended down the ages, allows close or distant views, always deeply penetrating, due to the balance of light and shade. Determined by slight adjustments, innumerable cows, and a few coppice hedges, from which solitary, timeless oaks emerge. A happy balance between the mass of woods and the pasture, the overall partitioning of the Limousin landscape.

The composition of the whole, while natural in appearance, turns out to be an artificial terrain, created by the engineer and the administrator. A rural space under scrutiny, managed by the agriculturalist and the forester. Looked at more closely, the smart dress of the Limousin is fraying at the edges. Whether historic or recent, a real depression is affecting the region and enriching it. A diversity unknown either in light or shade is appearing on features of the rural network: roadsides, river margins, edges and shoulders; but also in vast open spaces: heaths and peat bogs. These surfaces, although dispersed without any apparent order, offer in common a refuge for species driven out by the human management of forests and pastures. I have given the name "third landscape" to all those neglected territories, whose evident—and from now on necessary—function is to welcome those species that find no place elsewhere.

Can the name be considered as a real concept? For me, it offers a convenient means of analysis. Having achieved third position in the field of observation, perceived as the consequence of the existence of the primary and secondary—which dominate the situation—it seems obvious to afford it an equivalent or even superior weight, since it offers the only biological reservoir available, a genetic pool, a territory of the future.

By definition: an unresolved fragment of the planetary garden, the third landscape applies to all land above water level. It consists

[37] *Third landscape:* The term was considered by analogy with Third Estate and not Third World. Reference is made to the pamphlet of the Abbé Sieyès in January 1789: "What is the third estate?—Everything.—What has it been up until now in the political order?—Nothing.—What does it demand?—To become something."

of the sum of all urban and rural neglected spaces, *friches*, whether agricultural, industrial or touristic, but also all those spaces where, historically, management decisions have never existed: primary forests, high mountaintops, natural reserves, sacred spaces, and so on. It explains why shadow and light extend to populated and built-up areas, in the same way as to the empty and nonconstructed environment.

Whereas "the garden in movement" defines a management principle applied literally to the garden and even derived from it, the planetary garden and the third landscape appear as management principles of a political order, the responsibility of the community before that of the individual. Even if the individual is able to organize his daily life in accordance with these principles. Whatever the case, viewed as a whole, these positions are founded on observation and respect for the living. The projects that I am responsible for are not all necessarily driven by the same point of view. One cannot compare ideas, the generators of strictly localized projects, with concepts having a more general application. Intuition, impressions, analyses—idea-tools, objects of prime importance for setting in motion a method for elaborating the project, making the first marks of the design.

The idea emerges from an encounter with the site. It is intimately associated with it, and its only justification lies in this intimacy. Without this, it could be applied to anywhere in the territory. An idea becomes bad directly it ceases to be tied to a specific territorial feature. Exactly the opposite to the concept, which requires the whole mental and physical space in order to develop.

In the studios of the ENSH at Versailles, I try to develop a pedagogy by investing in the terrain a possible conceptual responsibility. I insist on a point that has been largely unexplored: in what way does the living organism, the inevitable object of our studies, manifest itself in each terrain, not as a simple point of discussion, but as the very subject of our work, the one thing that we must not overlook.

TEACHING

The living: the sum of all those beings capable of transformation,
from bacteria to man, bound together in a knot of relationships,
whether tight or loose, linking each part to the whole in a con-
stantly self-renewing dynamic.

On the outskirts of the city, on the edge of fields, the school accommodated about a hundred pupils. Outside, the revolt of the Sandi-
nista was confined to the dry forests of the north but a small group
resisted Jinotega, and in the cordillera Somoza, former chief of the
guardia, controlled the country and exploited it. Bled white, Nicara-
gua attempted to survive. For reasons connected with cooperation
agreements made between France and the countries of Latin America, I held the post of technical assessor at the agricultural college of
Matagalpa. Nobody ever explained to me what the role of an assessor was. Nevertheless, I can confirm that I held it in all conscience
for two years—it was my first teaching experience.

Since that day, by a trick of history, I have never ceased to
explore the function of teaching—part-time, tentatively, then more
and more confidently. First, to fill gaps in the institution, then over
the course of years—impossible to say exactly when—to elaborate
what with increased experience one believes to be a message.

The reasons for teaching are related to a desire for knowledge.
And, subsequently, to an overwhelming desire to share scraps of it.
But that would not be sufficient to explain what makes the teacher
dependent on those he teaches.

Something deeper, drawn from the depths of the unconscious,
no doubt very ancient, drives our need to transmit, to speak
urgently, as if nothing more important, not even feeding the body,
were more capable of allaying the anxiety that stems, no matter
what we do, from the very fact of being alive. This urge, apparent in
a thousand daily guises, brings people face to face defenseless. Circumstances, history, or simply the time of day designate on the one
hand, the adult said to be wise; on the other, the attentive child.
Roles unrelated to age. Someone knows what someone else does not

know. The teacher, taught, teaches afresh. He tries to penetrate the depths of ignorance—his, others—keeps his ears open, and begins over again. The edifice of knowledge is not built like a pyramid crowned with masters and scholars, but like a sketch of the cosmos in which particles develop through friction. The dispersed energies are reassembled, move away, become entangled in inextricable gibberish, become clearer through their own mirrors, lose their way, alight for a while on the horizon. Everywhere, the illusion of knowledge casts a halo round the knowledge of a thousand facts. Knowledge would be nothing without beliefs. Without anxieties.

Hope springs from this hesitation between certainties and questions.

This thing that we are talking about always comes from someone else, from the expression of another in us, working obscurely within each of us and transforming itself. Passed from one human being to another, from generation to generation, constantly elaborated, the message invents its vocabulary and leaves its rejects in the pages of dictionaries. The words of today are not those of yesterday. I believe in the heritage of ideas as a biological resource unique to the human race. Evolution.

Another reason for teaching. The emotional charges, magnetic spectra about which we know nothing, dispersed in the atmosphere of a classroom, a studio, or a lecture hall. Invisible, formidable bonds. Brief connections, tense exchanges, moments of defiance and boredom: to avoid them would be to destroy the fragile bridges that connect us, to compromise irrevocably the chances of communication, to plunge into disarray and silence those who, for one reason or another, are shouting.

At Matagalpa, I had ended up knowing the nicknames (*apodos*) of my pupils. Nothing strange for me about the use of *apodos*: dressing the other person in better clothing, identifying him by an infallible characteristic without actually confining him within it, naming in jest, taking it to the limit, being hurtful, going beyond the patronymic to turn it into the name of a warrior, knight, or commoner: accolades, country nicknames, familiar abuse, it all depends.

At the end of two years I had the exact list. Opposite each name, in alphabetical order, the *apodo*, discovered by chance and kept well secret: how should I make use of such an honor? What intimacy would have justified my using images, so tender, cruel, almost always just, invented by a clan from which I was by definition excluded? What is the invisible but permanent barrier that exists between the one who speaks the word and the one who receives it? Do messengers, which is what we teachers are, have a duty to retain a certain reserve? On what tightrope do we have to keep our balance? How do we remain aloof when the doors are thrown open to storms, laughter, stress? Invitations?

How to relieve oneself of the gift of those very people to whom one thought one was giving. Slowly and incoherently, an awareness was growing in me, throwing all former knowledge into confusion. The breakdown of all certainties. Eventually, I came face to face with this contradiction: trying to teach knowledge when I was learning to distrust it.

In an unpredictable, but nonetheless direct way, the garden in movement owes its existence to the principle of uncertainty recognized elsewhere: in the constantly changing world of human beings. No situation is considered definitive. What is implemented becomes established on condition that it is capable of modifying itself at any moment, easily, without any excessive expenditure of time and money. The energy endemic to it is considered essential, almost sacred. Any opposing energy risks destroying the natural dynamics, without even ensuring the success of the artificial dynamics put in place. Attempts to control render the space (the species) inflexible, robbing the garden of its character (similarly, the human being), transforming it into a prototype. The form of space that can be replicated—an ornamental and expensive commodity—distances itself from the garden in movement all the more, in that it suggests greater investment. These are indications that can be transposed to teaching. Does one want a robotic student from the École Nationale d'Administration or a thinking artist?

The respect due to all living things cannot be divided into nature on one side, humanity on the other. It is a matter of an inseparable whole, subject to the same laws of evolution. The same vulnerabilities.

At the agricultural school in Matagalpa, on the day before I left, I proceeded with the obligatory roll call, as on every Monday. That day I had decided to enter the pupils' world on an equal footing—not to assure them of an illusory equality between us, but to make them understand that the difference between teacher and taught is a matter of a moment, a look, and perhaps only of a decision. I picked up the list of nicknames.

At the sound of the first: amazement. I felt happy astonishment, hesitation, almost fear. At the second, there was a shudder like a wave passing through the rows. I didn't raise my eyes. They had understood. Indescribable happiness mixed with confusion. Impossible to erase from my memory the jumble of voices, the laughter, the noise of chairs, the disorder provoked by a public disclosure and the sharing of a game.

By the end of the list, the class in holiday spirit, were doubled up under the tables. I can still see Matiguas's machete lying on the desk beside his pencils, Evelio's revolver safe from the turmoil—to avoid an accident (almost all of them came armed)—the appearance of the bewildered director, forced to beat a retreat in the face of such complicity. Once again I can see that enjoyable disaster, the exposure of knowledge become suddenly unimportant. And then that musical litany, endless images that had become friends: Peor (the magpie), Loco (the fool), Vaca Holting (the cow Holting), Cara pelada (peeling face), Semaforo (the signal), Paludismo (malaria), Sicodelico (the psychedelic), Piruli taratamudo (the deaf-mute), Frentemono (monkey face), Timido (the timid), Siete pisos (seven floors), Bayardo (Bayard), Pelo de vaca (cow's hair).

Teaching is not the most important of my activities, yet it is the most important of my concerns. How to develop a pedagogy? Is it necessary? Can one restrict oneself to a method without running the risk of the teaching becoming fossilized?

Summoned by institutions short of teachers, I was compelled to hurriedly construct a teaching program, either nonexistent or embryonic at that point. Adapting the fragments of what I had been taught did not allow me to address the questions raised by the profession of landscape designer. Professional practice provided me with all the subject matter: the risks, the nature of the commissions, the contractors, the list of materials, and above all the incredible diversity of species capable of resolving the most basic problems and creating new ones.

I became interested in "the use of vegetation in the project" and in the project itself. The formal concept should not, on its own, take the place of a project. However, this was the general attitude in the schools. The influence of the plastic arts, seen from the reductive viewpoint of form alone, led the students to be satisfied with images. Art creates imagination before creating images. Each image contains a story. How can it be explored and transmitted?

For some obscure reason, probably due to the progressive separation of technical and artistic training, it was soon agreed that vegetation—reduced to a list—would not be part of the designer's brief until the final submission: a greenish sauerkraut applied to the architect's models at the last minute. At the School of Versailles, the students' spiraling loss of knowledge when it comes to plants stems from the fact that the studio work makes very little or no reference to it. The heterogeneity of recruitment creates a richness from which each individual profits, but it reduces to the lowest common denominator any knowledge of living organisms. The eagerly awaited teaching offered by the department of plant ecology manages to fill the immense gaps, and above all to develop an intelligence concerning the use of species.[38] It cannot replace a secondary school training that the schools calling themselves "superior"[39] refuse to offer. The

[38] At the time of writing, the department of plant ecology (at Versailles) is directed by Marc Rumelhart, assisted by Gabriel Chauvel and Olivier Jacqmin.

[39] *Écoles supérieures* are prestigious French high schools.—Translator's note.

scientific basis of the living world—the alphabet or notation if you like—remains unknown or poorly known. Hence, the use of limited and repetitive scales.

For a long time I had been struck by the preeminence of architecture in the landscape project. At a certain period—I estimate the peak of this tendency to have been in the 1980s—one might even have thought that it was frowned on, or simply considered old-fashioned, to talk about plants. This question, relegated to the works of fashionable women, could not concern the passionate designer in search of prestige. Any student quoting the name of a flower ran the risk of being treated as a gardener—a sort of insult—or even worse, as a florist. English gardens lack simplicity, tend to be collections. People sneered at them, forgetting that the smallest combination of species in a mixed border requires knowledge and know-how. They quoted Gertrude Jekyll, exercising her talent in the garden, her eyesight no longer allowing her to undertake her embroidery.[40] More often than not, no one was quoted, such was the prevailing culture's lack of references on a topic, from now on considered obsolete. A wave of professionals even attempted to usurp the title of architect by adding it to that of their own role. We owe it to the real architects, fastidious and jealous, for having saved the profession of "paysagiste" from a useless and pretentious confusion: they took official steps to ensure that the title that is their due did not end up insidiously validating other professions.

For the gardener, the question does not arise. Nobody lays claim to this title, a syndrome of a society that measures its superiority by the prevalence of intellectual professions, while manual professions remain at the bottom of the scale of values. An unfavorable context in which to learn about species, to observe their behavior, to

---

[40] Gertrude Jekyll (1843–1932), exploiting the contributions of William Robinson, worked to enrich combinations of flowers over the course of the year, using herbaceous plants. For many years she collaborated with the architect Edwin Lutyens.

anticipate a form of management based on that knowledge.[41] Yet, the project will only survive by being revisited over time, just as the gardener returns to his plot, observes nature's inventions and the development of his seeds, directs this energy, inflects it or adjusts to it. The garden is an observatory of time.

The landscape designer, at his own scale, observes the transformations of the landscape. From it he draws his argument for what it is to become. But he cannot speak seriously without taking into account the undeniable signs of a transformation that almost always forms part of the living domain. Forgetting compass and ruler, he measures the differences by the appearance and disappearance of species indicative of an environment. A structured catalogue of bio-indicators: providing, without any doubt, a serious tool for site analysis. Behind the forms of the landscape, there is life. Intervening in the forms, without paying attention to the deep-seated reasons behind their existence, is to endanger both the form and that which has caused it.

In his enthusiasm for decoding sites, Guillaume Geoffroy-Dechaume,[42] who recently left us, reached the innermost secrets of the landscape. Reading from one of his fragments, he would articulate the soil, the air, and the water, together with the calm or troubled past of a place. The flora tell the story, the fauna too. All you need to do is to bend down. Guillaume knew how to bend down. Certainly, he was one of the few landscape designers to keep a magnifying glass in his pocket, and to use it to surprise himself before surprising others. Together we shared the desire to understand, to

[41] The three-year practical course of gardening set up for students in the Potager du Roi (ENSP), on the initiative of the Department of Ecology, is considered as recreation with no impact on the curriculum, whereas it should be a fundamental exercise. Liliana Motta and Yves Gillen were directors of the operation at the time of writing.

[42] Guillaume Geoffroy-Dechaume collaborated with the Atelier Acanthe, which he directed until 2003. Among other numerous projects, he was the designer of the Parc du Chemin de l'Ile at Nanterre.

L'Île Derborence, Parc Henri Matisse, Lille, Gilles Clément, 2005.

question. A field of flowering plants, what better tribute? So many questions remain.

Some people carry the landscape within them, one might say "naturally." It is important to put quotation marks around this adjective borrowed from everyday speech: in reality, nothing is more firmly anchored in the *cultural* background of each one of us than the environment in which our eyes were first opened. This is the yardstick that defines the whole landscape. I am referring to those people who, from an early age, are led by a *natural* sensitivity to develop a subtle interpretation of the slightest alterations in their environment. People for whom everything is an indicator, the warmth of the air, the speed of the water current, the height of a tree, the weather . . . Such people involuntarily write the history of the landscape. We just have to get them to speak. Listen to them. Certain people, with reason, produce texts. From reading them, we can understand how the construction of a highway or the appearance of an exogenous plant on the shoulder—indicators of seemingly disparate value—are of equal force in changing the status quo, filling the memory with hope or anxiety, sometimes both. If the garden is an observatory of time, the gardener is inevitably an observer.

From these observers, I remember at least two enlightening pieces of work, both written by teachers. *Village-visage* by Jean Chatelut explains the transformations of a village where he was the caring mayor.[43] *Quand on avait tant de racines* by Adrienne Cazeilles describes the evolution of her region, Roussillon, over half a century.[44] I have met the authors and particularly, observed their characters. I keep up a constant dialogue with them. Positions to be taken, commitments. The division of public space, its infringement, its destruction, or more rarely its expansion. Perseverance, common sense, openness, are the characteristics of these minds. Saint-Benoît-du-Sault, south of Orléans in central France, Thuir in Roussillon in the Midi: two regions, two analyses, one conclusion: the landscape, as a social amenity, is the result of an administrative decision, and therefore political. At no point is it said that the predisposition of a region can be revealed simply by observing how it is maintained—a dangerously simplified reading—but no landscape solution can exist, even a temporary one, without decision-making power.

Anyone who addresses the question of landscape opens a political debate.

---

[43] Jean Chatelut, *Village-visage* (Saint-Benoît-du-Sault: Editions Payse, 2003).

[44] Adrienne Cazeilles, *Quand on avait tant de racines* (Canet-en-Roussillon: Editions Trabucaire, 2003).

Anyone who teaches the question of landscape is obliged to take a position.

The slogan written on the blackboard in '68—the tree is a capitalist—finds no response. That is because of the way the plea is formulated. But in its violence, the force of the plea demands that the tree be regarded differently. The status of social emblem has been added to that of ornament, biological organism, architectural feature. Again we have to come to an agreement. Is it a question of prestige or income? Marketed and patented, nature in its entirety might be regarded as a source of profit. In our role as teachers, we sometimes present the tree as a material, commercial object. But we prefer the living being, the one with which we share the earth, the light, the air and water. And hope.

"Do something, they're destroying your garden!" An alarming message from a journalist alerted by the rumor.

I was sufficiently alarmed, by the planning zeal of an authority barely aware of the consequences of its own "improvements": roads, paths, and parking lots. Modest developments, of little consequence to a passing observer, much more so for the inhabitant.

To be precise, a path runs along the side of my garden. What large-scale operation could involve the destruction of my own garden? From the window where the view to the west reaches the limits of the plot, I saw no sign of anything. Some insidious malevolence? A deadly cloud? Some invisible chemistry, more efficient than the fertilizers leached from the cultivated fields around—the residue of which inevitably reaches the lower slopes of the valley. What could be the reason for this sudden attack?

Paranoia. Mental aberration. Flashing images. It only takes a handful of seconds to imagine the worst. My informer insists: "You must do something, the bulldozers are there. They're tearing up everything."

On the horizon, no sign of a machine, not even the sound of an engine. It is not a question of my garden, but of one of those that are attributed to me because I am the concepteur. More than attributed, almost given to me, as if my intervention in the land—my creation—changed the status of the soil to the extent of altering its ownership. Newly elected, the mayor of the village where a garden is being razed does not share the ideas of his predecessor about the planning of public spaces. That is his right. The land is his property.

The garden disappeared. Like many others. Why shouldn't gardens disappear? Artificial devices, anonymous or signed, scattered over the earth: masterpieces, works of art, mediocre layouts, simple commodities. Ur, Babel, the gardens of Thebes, the terraces of Machu Pichu, the washhouse of Gometz-le-Chatel, the Perpignan train station: handfuls of stone, subject to erosion.

Every construction carries its own ruin within it. The visible, solid, piece of work, built from inert substances, confronts the attacks of time (or the bulldozers), with no other solution but to give in to them. But if discreet, organic, composed of living matter, it will discover ways of transforming itself without heading toward ruin. The garden is one such creation. Providing that in that place the gardener has not played at being an architect.

The machines destroyed the garden; nothing is left. The structures have disappeared. From memory, I knew that they were fragile. The land remains. Short of building a house (which is not the case), the worst that can happen—for me the best—is a *friche*, an abundance of species. It is not *my* garden, but it is not nothing. Certainly not a ruin. Just a landscape under construction.

The journalist would say, "The signature has disappeared." I say, "The garden has changed." Not through modesty about the piece of work, but because I know that where gardens are concerned, without exception, the author does not stand alone, in the form of an identifiable person, recognized by everyone as the artist. A single name is enough for the commissioners of the project, even if the reality is quite different. In every team the assistants work on numerous conceptual aspects—whether on the details or the broad outlines—in the course of the delicate phase of "tuning" ideas about the site. Sometimes, it is to the extent of influencing and reorienting the general concept. Who bothers to mention them?[45] Why don't they follow the rules of teamwork? Why is it that the list of credits—much shorter than in any other artistic production—never appears in the press articles? In the case of the Grande Ouche,[46] the garden bulldozed, modest in size, two of us were involved in conceiving and organizing the form of the spaces. Thierry Jour-d'heuil, my

---

[45] Only the professional journals such as *Urbanisme, architecture, paysage* take the trouble to mention all the participants.

[46] "Jardins de la Grande Ouche," Saint-Benoît-du-Sault, Indre, 1994.

fellow designer, was not informed about its destruction. No doubt, no one thought that the garden could be his as well.

The design studios understand this group mechanism. The initial idea, roughly sketched, intentionally clumsy (an artistic affectation by the "stars"), needs to be looked at again and revised. Sometimes rethought. Each person, depending on his or her particular skill, develops the idea, works on its refinement, variation, or level of precision. The architects, the designers, can assume that at the end of the chain an object ("the project") will appear, whose stable forms will be legible for a long time, corresponding to the design submitted. Hard landscaping withstands time. At least, a certain time. Not so the garden. It moves with it.

No sooner installed, it begins to unfold. The designed forms are confirmed. But time works at unraveling the design. It outmaneuvers the knowledgeable speculations, the fine anticipations. It transforms certainties into daily questions and knowledge into a mass of data, juxtaposed according to their academic logic. Inert. Turned on their head once exposed. Indispensable and inadequate.

Precision and appropriateness: the gardener defines the species supposed to control the space, to give it a style. He repeats them wherever necessary. Plants them, watches over them. With them alone, he cannot hope for a garden. Only a geometry. In addition, he has to accept plants that are fragile, "vagabond," diverse, and abundant. Accept seeing them escape or establish themselves: invade the ground, mark the landscape to the extent of giving it their signature. Giving it a supplementary identity. Might they be able to efface the artificial framework of the planted vegetation?

Teamwork, the work of time, multiply the opportunities for partition. The gardener, inheritor of the project, accommodates himself with time as best he can. He diversifies his methods and forms. Sometimes simplifies them. It would be naive to imagine that he does not put his mark on the territory and, perhaps, offensive to prevent him.

The three principal reasons for the "partition"—the weakening of the concept, the ascendancy of nature, the interpretation of the gardener—contribute to the constant evolution of the forms supposed to represent the idea.

There are ways of trying to project the image into the future. The oldest put to the test: to harden the forms to the extent that they resemble the art of building. From the Pompeian peristyle to the great classical gardens, there is no lack of examples. In that case, the lasting quality of the work is achieved by constantly updating the plan on the ground. The plan, a reassuring tool. Those who support historic reconstruction refer to it, hand laid on the archives as if on the Bible, swearing by rigor and accuracy.

Another more contemporary procedure, borrowed from the legal complexities coming from the other side of the Atlantic, is to protect the work as the inalienable creation of the artist. The garden, considered as a work of art, registered as such in the inscrutable memory of whatever fund is delegated to this job, cannot be modified without the authorization of its author. That is how the Jardins de l'Imaginaire at Terrasson in the Dordogne operate. The author Kathryn Gustafson has insisted on this contract as landscape-artist.

The garden attracts artists. The scope of the subject, the infinite number of implementations possible, the questions raised by the relationship between man and nature, all conspire to seduce anyone attempting to reinterpret the world. Certain artists have not been deceived as to the frailty of a scenography based on living material. The mold introduced by Michel Blazy, the evaporation introduced as a driving force by Jean-Luc Brisson: interventions in which the outcome, neither its form nor appearance being predictable, is left to the initiative of the environment. Where the very timing of the outcome—the image of a garden—is not fixed in advance. Who can tell at what opportune moment the garden will be completed? Who can assess the adult age of a garden? Such notions belong to the list of those works controlled from beginning to end, like a sculpture. If the garden were a sculpture—certain ones make reference to it, but are they gardens?—it would be necessary to return to it every day, and faced with the inanity of such practices, never lose faith.

Territory of uncertainty—of our own uncertainty—the garden transforms our derisory gestures into sacred moments. Should it be "in movement," this garden, then these rare moments that we need so much would multiply. Surprises. So much precision in the position of a branch newly discovered. Although hanging there for a long time, it suddenly reveals a view, or masks it in an unexpected way. The subtlety of silver foliage in front of an elusive blue, a columbine, forgotten against the excessively hard green of ordinary grass. The lungwort, the variety will have to be identified. The carpet of stitchwort is reaching the undergrowth, covering up the cyclamen bulbs. Their foliage invisible at this time of year. Where have they spread now? Wait for the autumn. The ants transport their tiny seeds. They go everywhere. Right to the end of the coppice, I think. The blackthorn in the hedge can be shorn like a sheep. It has the same shape. Allowed to grow too high, it would hide the fig tree at the end of the orchard. No figs on it, it would be better off on the slope in the sun. The lotus has only produced one flower, the water beetles don't leave so much as a frog's egg in the pool, they are assassins. Remove the shepherd's purse, is that really necessary? An annual spurge is spreading, along with the sun spurge. And the caper spurge, and again the woodland spurge at the edge of the garden. This morning I am going to weed there, around the ice plants, the fumitory is getting tangled up with the sage, it's looking for the light, it will dry up in June, we'll see later. Yes, the meadow grass is interrupting the line of orangey wood betony, it's tangled up with the gray fescue and hardy oats. By scratching the soil a little underneath the hellebore, a seed or two might soon germinate, they can't bear waiting. The lady's mantle is taking over a good square meter under the Moroccan broom and the white heather is suffering from the shade, the smell of citronella comes from the ground, I must have broken off a geranium rhyzome, the mole has lifted the bulbs of the gladioli, very small ones, wild, pink, and delicate, like those that

grow in the *friches* of the Maghreb and on the hillsides of Nice, that's
going too far. Here I could sow nasturtiums over the garlic bulbs.
Where did I plant the garlic? The voles are eating the tulips, I knew it.
Hope remains that one fine day, some species will dissuade them, or
else they will be put off them by overeating. That oak at the bottom
will provide wood for the winter, it shuts off the valley, devours the
light, the blackbirds have harvested the blueberries, a young plant of
Chinese pinks that I'm being given, put it in front of the oval hole like
a large dish? That would take away the sun from the green lizards, to
the left of the ceanothus perhaps, or better still: in the company of the
Nozomi rose bushes that are disappearing before my very eyes. Competition from the bamboo, a clump of ten giant canes, of which at
least three would make good stakes for the tomatoes, the old chestnut
posts are losing their heads and holding out, but their numbers

dwindle year by year, I should take them for dry wood and burn them without further thought, or use them to lever up a stone, or do nothing at all and lose them in a wood. The apple tree is resprouting from the fallen trunk, must cut its poorly irrigated top. Prepare a prop for the ironwood tree. It blocks the path lower down. Free the golden saxifrage, invaded by deadnettle and reed, it's damp there, fix the rabbit wire on the beech steps, get rid of the brambles except in the hedge. Avoid the unwieldy Gaultheria on the edge of the stream, a wren is nesting there, I can't do anything about it. The leaves of the gunnera, too heavy. The stump is breaking up, it's going to produce little ones. The Californian poppies are seeding in the avenue—avenue, that's a grand word—like the carrots, parsley, and other fine seeds, they like soils that are well packed and drain well, the water does not stay around, it disappears goodness knows where. Impossible to know where the water disappears to, but I'm convinced that it carries with it the small amount of manure produced by the Dutch cows (why Dutch?), little compressed grains, without any excessive smell, placed a foot apart, to dissolve, so as to be more easily eaten. Biscuits. There is no drain across the path. Pile up some stones, reverse the gutter on the shed, it's flowing backward, I won't prune the Nelly Moser clematis growing on the red cornus, I'm a bit ashamed of its oversize flowers. Higher up, dispersed, there'll be just a hint of them. The clipped privet ball finally looks like a real shrub, I don't prune it any more, in the heat wave the passionflower has managed five flowers, the wisteria none, it's sulking.[47]

---

[47] Botanical names of all the plants quoted in this passage, in order of appearance: *Aquilegia vulgaris, Pulmonaria sacharata, Stellaria holostea, Cyclamen repandum, Prunus spinosa, Ficus carica, Nelumbo nucifera, Capsella bursa-pastoris, Euphorbia heliscopia, Euphorbia lathyris, Euphorbia sylvestris, Sedum spectabile, Corydalis claviculata, Salvia officinalis, Poa annua, Helleborus orientalis, Alchemilla mollis, Cytisus battandieri, Erica X 'Silberschmelze', Geranium maccrorhyzum, Gladiolus byzantinus, Tropaeolum majus, Allium spherocephalon, Tulipa tarda, Vaccinium corymbosum, Ceanothus thyrsiflorus, Rosa 'Nozomi', Bambusa sulfurea, Parrotia persica, Chrysoplenium oppositifolium, Lamium maculatum, Glyceria aquatica, Gaultheria shallon, Gunnera manicata, Eschscholtzia californica, Cornus sanguinea, Passiflora caerulea, Wisteria sinensis.*

Meadow raft at La Vallée.

Perhaps a more shaded terrace for the summer? Change the location of the compost, it's not that simple, the piles of wood lined up on the side. A shelter for the logs, instead of this tarpaulin. In the meadow, an awning to screen the sun. Placed on the brow of the field, the raised ground where the deer come, a seat or even a reclining chair, to sit or lie, to observe the flowers in succession, the tiny insects, little clouds, aircraft, hesitant butterflies, speedy wasps; the birds in the distance, crows as always, magpie shrikes (*lanius*) in the hedge behind; sometimes a buzzard and its cry, sometimes a kite, never the two together. The oaks enclose the field to the south. The horizon glimpsed in spring and autumn. Perhaps observing nothing. Unless we should look within? What do we see there, I put the question, is it the mirror of an internal garden, walled in accordance with rules that remain unknown, which might reveal itself from time to time, when time permits, when we make the decision, or rather, when a space within us confronts this risk, the sublime desire: to look?

And if that alone were the signature?

If, in reality, doing and knowing were to be eclipsed by amazement? The state of things unveiled—or semi-veiled—offered undisguised, and translated into words (sometimes into images)—seeing is nothing unless expressed—could it belong to each one of us, like a personal treasure? A unique way of reading and speaking? Of writing, and signing.

We are imbued with everything that surrounds us, imbued with others, how could it be otherwise? How can we pretend to the ownership—of an idea, of an object—when everything around us is aglitter with calculated pressures? How can we know what drives us?

I have asked myself these questions, because people have asked me. If not, why worry? A legitimate anxiety: how can you guarantee the work, say the sponsors, and how can I, your sponsor, be assured that the image will be preserved? Haven't I bought it from you? Haven't I the right to hope that it will be maintained in the future?

In my own garden, I, the gardener, wouldn't be able to predict its exact form tomorrow. Only the moment exists. So how to guarantee the future of a distant territory where one doesn't even know the birds? Gardeners succeed one another, those who give orders don't stay around for long; in place of the image—modified by one and all—could the spirit be made to endure?

Of all the gardens that I am responsible for, only the Domaine du Rayol has been the subject of a followup.[48] No doubt, the Conserva-

---

[48] The Domaine du Rayol, on the Mediterranean coast about twenty kilometers west of Saint-Tropez, covers twenty hectares and is managed by the French government's Conservatoire du Littoral. Designed by Clément as a series of Mediterranean gardens whose vegetation draws on the different countries within that climate, the landscape reflects the philosophy of the garden in movement and the third landscape.—Translator's note.

toire du Littoral is aware of the frailty of what it manages? No doubt, it takes into account the inability of architecture alone to sustain the project. It knows that living material cannot be controlled and that the signature, if such a thing exists, is the result of a lack of precision, a feeling—the spirit of the place, actually—rather than legible, perfect—disenchanted—forms.

Perhaps the gardener is not someone who makes forms survive over time, but over time, if possible, ensures that enchantment survives.

We must try.